Teaching Montessori Science

9 Practical Strategies to Engage Children in Hands-on STEAM Activities

Jackie Grundberg

NOVA TERRA

Publishing

Book Cover Design: GetCovers
Layout Design: Arvin Trinidad
Editor: Noelle Chow

ISBN Ebook: 979-8-9902346-0-4

ISBN Print: 979-8-9902346-1-1

Dedicated to Brent, Jonathan and Katherine

A Free Gift for Our Readers

Receive a free lesson plan on the Skeletal System.
A full written and video lesson plan with materials, procedures, presentation cards, Montessori 5 part cards, and follow-up activities.

https://www.backpacksciences.com/skeletal

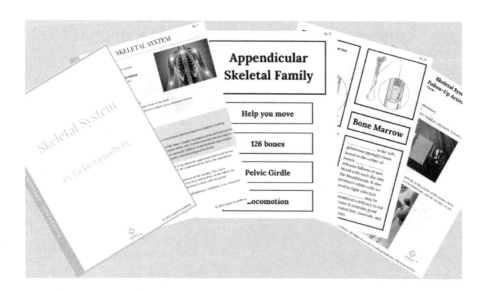

Introduction

*"Tell me, and I forget. Show me, and I remember.
Involve me, and I understand."*

This timeless adage, attributed to both Confucius and Benjamin Franklin, encapsulates the essence of effective learning. It emphasizes the importance of active engagement in the educational process – a principle that lies at the heart of Montessori education and resonates deeply with our approach to teaching science.

In our quest to inspire children and cultivate a passion for scientific inquiry, we recognize the transformative power of hands-on learning experiences. By encouraging children to actively participate in experiments, explorations, and real-world applications of scientific concepts, we pave the way for profound understanding and lasting retention.

As a former wildlife biologist, I have conducted hands-on research in the field, fostering a deep appreciation for experiential learning. Armed with a Master of Education in Educational Technology and Montessori credentials to teach 6- to 12-year-old children, I've spent over 25 years developing my expertise at the intersection of science and education. Additionally, my state certifications to teach high school biology and middle school underscore my commitment to delivering quality STEAM education at all levels. As the founder of Backpack Sciences, a thriving business providing science curriculum support and coaching to teachers and parents, I have leveraged my expertise to empower educators and families alike. Having been invited to speak at numerous Montessori conferences, I have shared insights and best practices with educators worldwide. Through this diverse blend of experiences, I bring a wealth of practical knowledge and academic rigor to empower educators in implementing the strategies outlined in this book, bridging the gap between theory and impactful classroom practice.

In this book, *Teaching Montessori Science: 9 Practical Strategies to Engage Children in Hands-On STEAM Activities*, we embark on a journey to explore innovative approaches to science education. Each chapter is dedicated to uncovering practical strategies, with detailed examples, that empower educators to create dynamic learning environments where every child can thrive.

What sets *Teaching Montessori Science: 9 Practical Strategies to Engage Children in Hands-On STEAM Activities*' apart is its dual emphasis on providing the 'why' behind each strategy and offering concrete, actionable examples for implementation. Unlike traditional educational guides, this book not only outlines the importance of hands-on, interdisciplinary science education but also equips educators with specific techniques and real-life scenarios to effectively integrate these strategies into their classrooms, ensuring a seamless transition from theory to practice.

Join us as we dive into the benefits of hands-on, interactive science lessons, explore the importance of inclusivity and cultural sensitivity in science education, and harness the power of storytelling to captivate young minds. Together, we'll discover how to leverage choice, real-world issues, technology, observation, and community-building to enrich the learning experience and inspire a new generation of scientific thinkers.

Table of
Contents

Chapter 01

THE BENEFITS OF HANDS-ON, INTERACTIVE SCIENCE LESSONS

Bringing science to life

Movement of the hand is essential.
Little children revealed that the
development of the mind is
stimulated by the movement of the
hands. The hand is the instrument
of the intelligence. The child needs
to manipulate objects and to gain
experience by touching
and handling.

Dr. Maria Montessori
The 1946
London Lectures

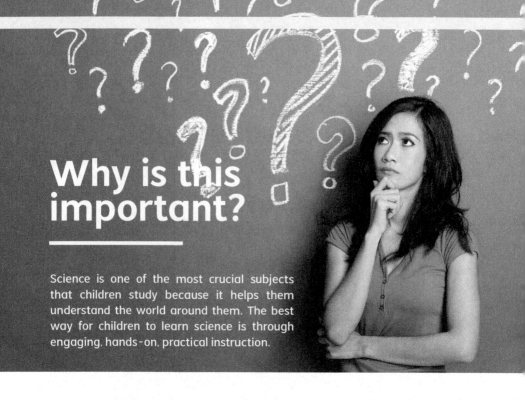

Why is this important?

Science is one of the most crucial subjects that children study because it helps them understand the world around them. The best way for children to learn science is through engaging, hands-on, practical instruction.

Kids become more engaged and interested in what they are learning when they participate in experiential activities. By developing their senses, children can learn more about the world around them. For example, they can go outside to explore nature or construct, study, and dismantle models. When we use strategies like these, children are unaware of the complexity of the scientific concept. It just becomes fun.

When children actively participate in activities that involve exploring, constructing, experimenting and using their senses, there is a sense of joy or a feeling of "play." For example, an educator might have the children investigate a simple chemical reaction of baking soda and vinegar. When baking soda (sodium bicarbonate) is mixed with vinegar (acetic acid), a chemical reaction occurs that produces carbon dioxide gas, water, and sodium acetate. The carbon dioxide gas is released as bubbles and creates pressure within the mixture, causing it to erupt in a foamy eruption.

Most likely, as the children experience the foamy eruption, they are probably not initially thinking about the chemical reaction. They are having a great time seeing the outcome of bubbles overflowing the volcano. This fun and exciting example can engage children and get them excited to learn more about volcanoes, geology, and the properties of gas.

A recent study demonstrated that activities that involve hands-on, interactive, and experiential learning are beneficial for teaching science to elementary school students. In 2019, Eshach and Fried reviewed 33 studies and discovered that students were more interested in science and learned more about it through hands-on and experiential activities. For younger students, these advantages were even more pronounced.

BACKPACK
SCIENCES

Hands-on activities were found to be effective for promoting science learning in young children in another study by Bartos and colleagues (2019), particularly when the activities were designed to encourage inquiry-based learning and experimentation. Both children's attitudes toward science and their comprehension of scientific concepts were enhanced by these activities, according to the researchers.

Experiential learning helps children comprehend the ideas they are learning. The topic takes on a deeper meaning and stays with them longer when they encounter the scientific idea in its natural setting. Field trips to museums, nature centers, and conversations with science experts are all examples of experiential learning. Through activities like these, children can better understand the relevance of science to everyday life. Children learn to love science, practice asking questions, and develop a sense of wonder about the world around them. In Falk and Dierking's (2000) study, children who participated in science-related field trips and other experiential learning activities demonstrated increased knowledge and interest in science.

These findings are consistent with the constructivist theory of learning, which suggests that students learn best when they actively participate in their own learning (Driver et al, 1994). This theory holds that students construct their own knowledge through their experiences and interactions with the world around them. Hands-on, interactive, and experiential activities provide opportunities for children to construct their own understanding of science and make connections between their experiences and scientific concepts.

As educators, we want to include interactive activities, such as group projects and discussion, because they play a crucial role in helping children understand science. When children work together and share their discoveries, reflections, and ideas for next steps, they can build on each other's understanding and gain a deeper appreciation for the subject. Another benefit to interactive activities is the development of the child's social skills - for example, communication, listening, cooperation, and questioning.

An 'F' in Science!

I can't say I always loved science. In 5th grade, I received an "F" in science class. When I think about my experience of science in elementary school, I can't remember anything... not even ONE thing.

What I do remember is how angry my parents were. My mom was a nurse, and my dad was an engineer. Later on, my sisters became a medical doctor and an engineer-so as you can imagine, at that moment, I was the odd man out.

When I got to middle school, it was a different world. My school was right on the edge of a cliff overlooking the Pacific Ocean. Across the parking lot of the school was a footpath that took you down to the beach. Right at the bottom of the path, before you got to the sandy beach, there were jagged rocks.

One science class, we ventured down and started exploring the tide pools at low tide. Oh.... the things I remember seeing: starfish, brittle stars, and small fish swimming around colorful sea anemones. One time I even found a shark egg case! I saved that until I left for college.

Throughout middle school, we went down to the water several times. It was an easy field trip for the science teachers, and I loved every trip. No trip was the same, and it was because of those experiences that I wanted to be a marine biologist. In high school, I took a summer college course at Scripps Institution of Oceanography at the University of California, San Diego.

This was just the start of why I'm so passionate about encouraging teachers to include hands-on, interactive, and experiential learning in science.

BACKPACK SCIENCES

5 Tips to Bring Science To Life

By using materials and manipulatives, we allow children to engage, explore and experiment. Children do this by participating in interactive activities, collecting data and observations, and fostering a sense of ownership. Students can gain a deeper connection to the materials and retain their knowledge more effectively.

01 Use hands-on materials and manipulatives to help children visualize and understand science concepts

In order for children to visualize and understand scientific concepts, we need to include hands-on materials and manipulatives. This should start with the teacher's introduction and presentation of the concept. As teachers, we want to make an "impression" on the child's mind.

This approach allows students to physically engage with the material, making it more memorable and easier to retain. The "impression" doesn't have to be anything spectacular. It can be objects that have some relationship to the science concept, or it can be a story (which will be discussed in another chapter).

Hands-on materials and manipulatives are also a must once the presentation is completed and it's time for the children to explore the scientific concept further. Allowing students to explore and experiment with hands-on materials fosters their natural curiosity and creativity.

For example, if you're teaching about roots, you'll want to bring in examples of roots. This can be the entire carrot-from the stems to the whole carrot (taproot). If you soak the carrot root body in water for a few days, you might be able to get some smaller lateral roots emerging from the sides.

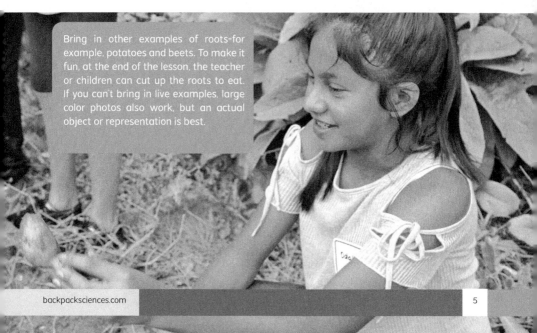

Bring in other examples of roots-for example, potatoes and beets. To make it fun, at the end of the lesson, the teacher or children can cut up the roots to eat. If you can't bring in live examples, large color photos also work, but an actual object or representation is best.

Encourage students to explore and experiment with hands-on materials

Here are six examples of how you can use hands-on materials and manipulatives to help children visualize and understand science concepts.

01 Building models of molecules

A Present to the children the basic concepts of atoms, molecules, and chemical reactions.

B Provide the children with a set of materials to build models of molecules. This may include a commercially purchased set or items purchased at a grocery store-for example, toothpicks and marshmallows, beads, or colored paper.

C Allow the students to research the properties and structure of the molecules they choose to build. Examples include water, glucose, and carbon dioxide.

D During the research, the children should they take note of the number and arrangement of atoms.

E Have the children build the models of molecules. Compare their structures and properties to those other students created. What do they notice is similar or different?

F Encourage the children to explain how chemical reactions occur, based on the interaction of the atoms and molecules.

G Conclude the activity by summarizing the main observations and conclusions, and discuss the relevance of chemical reactions in everyday life.

02 Constructing circuits

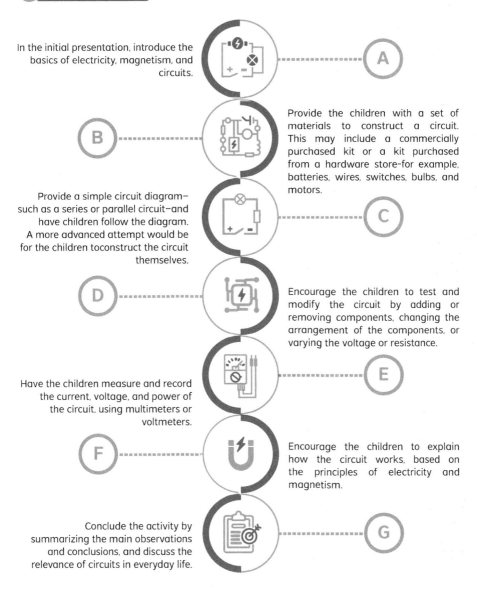

In the initial presentation, introduce the basics of electricity, magnetism, and circuits.

A

B

Provide the children with a set of materials to construct a circuit. This may include a commercially purchased kit or a kit purchased from a hardware store-for example, batteries, wires, switches, bulbs, and motors.

Provide a simple circuit diagram— such as a series or parallel circuit—and have children follow the diagram. A more advanced attempt would be for the children toconstruct the circuit themselves.

C

D

Encourage the children to test and modify the circuit by adding or removing components, changing the arrangement of the components, or varying the voltage or resistance.

Have the children measure and record the current, voltage, and power of the circuit, using multimeters or voltmeters.

E

F

Encourage the children to explain how the circuit works, based on the principles of electricity and magnetism.

Conclude the activity by summarizing the main observations and conclusions, and discuss the relevance of circuits in everyday life.

G

03 **Using Play-Doh or clay to model geological processes and landform development**

A Present to the children the basics of geology, such as plate tectonics, erosion, deposition, and weathering.

B Brainstorm with the children what materials they can use to model geological processes, such as Play-Doh or clay, sand, water, or rocks.

C Have the children research the specific geological processes and landforms they'd like to model, such as earthquakes, valleys, mountains, or beaches.

D Encourage the children to model the geological processes and landforms by including the main features and processes, and explain how they are formed and changed over time.

E Encourage the children to observe and compare the models of different students and discuss the similarities and differences.

04 **Modeling simple machines**

A Start the lesson by talking about the basics of force, energy, and physics.

B Show real-life examples of levers, pulleys, wheels, axles, and other simple tools.

C After the lesson, give the children time to manipulate and test the simple machines on their own and watch how they work and interact. They should pay attention to what makes each simple machine different.

D If possible, have the children use rulers, scales, or spring scales to measure and write down the amount of force put in, the amount of force put out, and the distance. (This is a little more advanced.)

E After the children have had a chance to interact with the simple machines, ask them to explain how they change the force and direction of both the input force and the output force.

F Another good follow-up task is for the children to look around at their surroundings and notice all the simple machines they use every day.

BACKPACK
SCIENCES

05 Creating a greenhouse environment

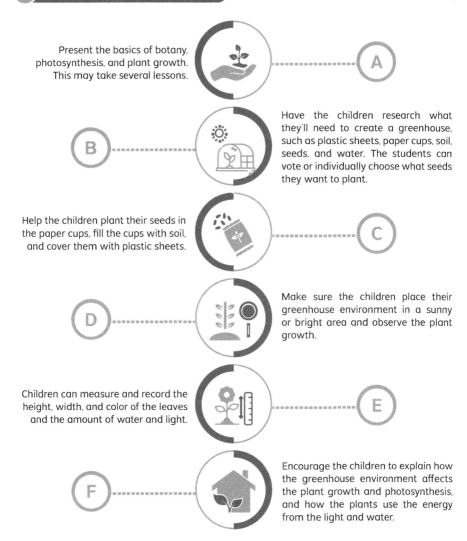

Present the basics of botany, photosynthesis, and plant growth. This may take several lessons.

A

B

Have the children research what they'll need to create a greenhouse, such as plastic sheets, paper cups, soil, seeds, and water. The students can vote or individually choose what seeds they want to plant.

Help the children plant their seeds in the paper cups, fill the cups with soil, and cover them with plastic sheets.

C

D

Make sure the children place their greenhouse environment in a sunny or bright area and observe the plant growth.

Children can measure and record the height, width, and color of the leaves and the amount of water and light.

E

F

Encourage the children to explain how the greenhouse environment affects the plant growth and photosynthesis, and how the plants use the energy from the light and water.

A Present the basics of thermodynamics, heat, and energy transfer.

B Set up an experimental area with a set of materials to measure the rate of evaporation and condensation. Materials include thermometers, cups, liquids, and towels (paper or cloth).

C Have the children choose two different liquids-for example, water, alcohol, or oil-and pour them into two separate cups. (Later, you can discuss whether cups made with different materials would make a difference.)

D Have the children wrap the cups with paper or cloth towels, and place them in a sunny or warm area.

E Have the children use their scientific notebooks to keep records of the temperature and volume of the liquids at regular intervals-for example, every 5 or 10 minutes.

F After some comparison, analysis, and reflection, encourage the children to explain how the rate of evaporation and condensation is affected by the temperature, volume, and composition of the liquids.

03 Provide opportunities for children to participate in interactive activities that involve movement

A wonderful way to incorporate a science concept is to allow children's bodies to learn through movement. Research has identified that adding movement and exercise in study periods can boost verbal memory, thinking, and learning. This movement can help recall and understand new vocabulary.

Movement includes small gestures or motions - for example, a hand signal for a vocabulary word. A wonderful example Larry Ferlazzoo gives is for the word "weathering." He makes a chopping motion with one hand, representing wind or water chipping away at a rock. He then describes how he uses the motion of one hand moving like a wave, representing the movement of sediments from one place to another for "erosion." Each time he uses the vocabulary word, whether in a presentation or just in conversation, he uses the gesture or movement as well.

Other activities that involve movement are simulations and role-playing. Here are three examples:

01 Simulating the spread of disease

A Present to the children the basics of how diseases spread and how they are controlled.

B Divide the children into two groups: "carriers" and "uninfected."

C Assign each carrier a specific disease.

D Provide each uninfected with a set of cards representing their health status.

E Have the carriers move around the room and "infect" the patients by touching them or exchanging cards.

F Have the patients mark their health status and track the spread of the disease.

G After a few rounds, discuss with the children the effectiveness of different control measures, such as quarantine, vaccination, and hygiene practices.

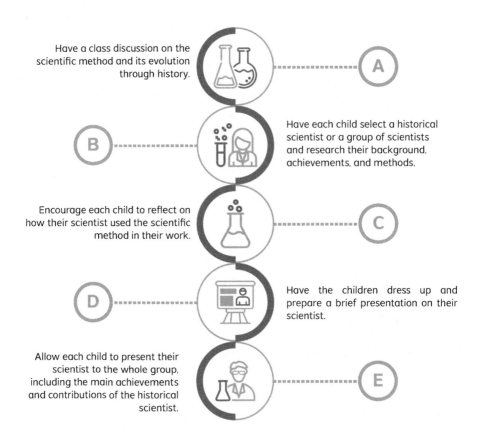

Have a class discussion on the scientific method and its evolution through history.

A

B

Have each child select a historical scientist or a group of scientists and research their background, achievements, and methods.

Encourage each child to reflect on how their scientist used the scientific method in their work.

C

D

Have the children dress up and prepare a brief presentation on their scientist.

Allow each child to present their scientist to the whole group, including the main achievements and contributions of the historical scientist.

E

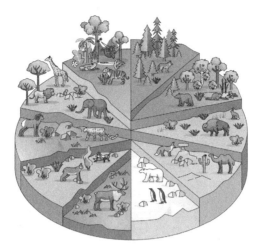

03 Creating a virtual ecosystem

A Present to the children the basics of an ecosystem, such as the relationships between predators and prey, food chains, and ecological balance.

B Divide the students into groups and allow each group to choose a specific ecosystem, such as a rainforest, desert, grassland, or tundra.

C Have the children research the specific ecosystem and its inhabitants, including flora (plants) and fauna (animals).

D Encourage each group to present their ecosystem in a way that will represent the ecosystem. This may include a diorama, a painting, a model, fabric, or figurines.

E Have the children present their ecosystems to the group, including the different species and their relationships.

F Encourage the children to add, remove, or alter the elements of the ecosystem and observe the changes in the predator-prey relationships and ecological balance.

G Have the children conclude their presentations by summarizing the main observations and conclusions of their ecosystem. Discuss the importance.

Have children collect data and make observations

Children can see how science ideas are applied in the real world by gathering data and making observations themselves. This is how we teach children to slow down and pay attention to nuances that they might otherwise overlook. A good example is an ant trail on the sidewalk. On a daily basis, this will go unnoticed, but if you pause and pay attention, you will notice an amazing system of cooperation and communication.

When we show children how to take effective notes, they learn to pay attention to nuances, examine facts, and pose questions.

This enhances scientific literacy growth, motivation and engagement, and critical thinking and problem-solving abilities.

Here are some examples of activities:

01 **Measuring and tracking changes in temperature and atmospheric pressure during a weather study**

A Obtain thermometers and barometers. You may need to refresh children's memories or teach them how to read these.

B Have the children place thermometers in various locations, both inside and outside their learning environment

C Have the children create a schedule to record the temperature and atmospheric pressures several times a day over a period of time.

D The children may choose different ways to visualize the data-for example, different types of graphs and charts.

E Ask the children to notice and discuss the patterns and changes they observed in the temperature and atmospheric pressure.

02 Conducting a soil survey to analyze the composition and structure of different soil types

Present basic soil importance, properties and the different soil types.

A

B

Have the children collect samples of different areas-for example, a park, beach, garden, forest, etc.

Give the children hand lenses or microscopes to examine the soil particles.

C

D

Have the children take notes and make drawings of the different textures and structures."

Have the children create a table or chart to organize their observations.

E

F

Discuss the different soil types observed and the characteristics that make each unique.

03 Observing an organism's behavior (for example, a pet or known anthill) over a period of time

A Have the children select a habitat-for example, a garden, a park, or a classroom.

B Make sure the children have a scientific notebook or observation journal.

C Visit the habitat several times over a period of time so the children can observe and record the behavior of the organism.

D Ask the children to make sketches and notes of the organism's behavior, including behavior (including hypotheses, or guesses, of why the organism is behaving that way).

E Encourage the children to create a chart or table to organize their observations.

F As a group, discuss patterns and behaviors observed.

05 Foster a sense of ownership and responsibility for hands-on materials by allowing children to take care of and use them regularly

One crucial component of a hands-on approach to teaching science is fostering a sense of ownership and responsibility. I encourage children to be responsible for taking care of the science materials and equipment. This may include purchasing, cleaning, and maintenance.

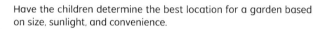

01 Having the children choose a part of the class or home garden and track its growth and development

A Have the children determine the best location for a garden based on size, sunlight, and convenience.

B Provide the children with age- and size-appropriate garden tools, gloves and seeds. Choose "real" tools, not child-like, play versions of the real tools.

C Map out and record which area each child is responsible for.

D Plant seeds or seedlings.

E Children can record the growth and development of their plants in their scientific notebooks.

02 Providing children with a personal set of science lab equipment and asking them to keep it organized, clean, and functioning

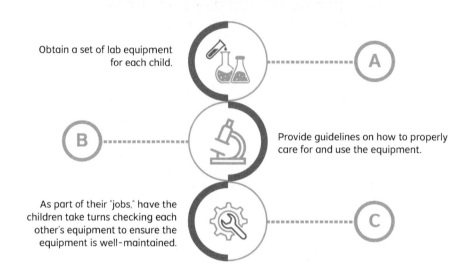

Obtain a set of lab equipment for each child.

A

B

Provide guidelines on how to properly care for and use the equipment.

As part of their "jobs," have the children take turns checking each other's equipment to ensure the equipment is well-maintained.

C

03 Encouraging children to create and maintain an environment for a living organism-for example, an aquarium or pet (guinea pig or gerbil)

A Have the children research and create a list of all the supplies needed-for example, an aquarium, water, rocks, plants, fish, food, etc.

B Have the children create a checklist and guidelines of how to properly care for and maintain the aquarium.

C Create a sign-up or rotation list that dictates who takes care of the aquarium.

D Ask the children to keep track of the aquarium's care and maintenance in their scientific notebooks, and help them track the growth and behavior of the aquatic life in the aquarium. This may become crucial documentation if any issues arise.

E This will provide opportunities for the children to observe, measure, and compare the growth and behavior of different aquatic species.

BACKPACK
SCIENCES

COMMON CHALLENGES IN INCORPORATING HANDS-ON, INTERACTIVE, EXPERIENTIAL LESSONS

When incorporating hands-on, interactive science lessons, education professionals frequently encounter a number of difficulties that can make it challenging to effectively engage students in the learning process. Three of the most frequent problems are a lack of resources, time restraints, and issues with pupil behavior management. Teachers can successfully integrate practical lessons into their curricula, increasing student engagement and comprehension, by understanding these difficulties and looking into potential solutions.

I DON'T HAVE ALL OF THE RESOURCES AND SUPPLIES.

Many schools and classrooms may not have access to the supplies and materials necessary to create engaging, hands-on science lessons.

Solution:

Look for resources, companies, and lesson plans that use everyday household items-for example, paper cups and rubber bands-to create low-cost, hands-on experiments that still effectively teach scientific concepts. Utilize local resources such as parks, nature centers, museums, and community organizations that provide materials and resources.

I DON'T HAVE ANY TIME.

It can be challenging for teachers to incorporate interactive, hands-on science classes due to the short class periods and the extensive curriculum. This also applies to the planning that goes into creating and preparing these kinds of classes.

Solution:

To make the most of your time and to create interdisciplinary classes, try incorporating practical activities into other subjects, like math and language arts. (Read the chapter on interdisciplinary lessons for specific examples.) Students' interest in science can also be piqued by adding quick, interactive activities to regular lessons, like interactive demos or observations.

THE KIDS GET SO EXCITED WITH HANDS-ON ACTIVITIES THAT IT'S DIFFICULT TO MANAGE THEIR BEHAVIOR.

Activities that require hands-on participation can be messy and can disrupt the learning atmosphere.

Solution:

To reduce potential disruptions during hands-on activities, establish clear rules and procedures early on. This should include letting kids assume responsibility for their learning space. If you can, go outside to make the commotion more bearable for those nearby.

5 QUICK FIXES

Five small, doable suggestions that can lead to quick results for educators or homeschooling parents looking to implement hands-on, interactive examples into their science lessons

1. START WITH SIMPLE HANDS-ON ACTIVITIES

For example, you can use real fruits and vegetables when teaching botany or kitchen supplies to teach about chemical reactions.

2. USE MANIPULATIVES IN YOUR LESSONS

Manipulatives are physical objects that children can touch and move around to help them better understand a scientific concept. Examples include a model of a skeleton or different plant leaves.

3. ENCOURAGE EXPERIMENTATION

Encourage children to experiment and try things out themselves. This can involve simple activities, such as mixing different substances to see what happens or conducting a simple science experiment.

4. MAKE CONNECTIONS TO REAL LIFE

Hands-on science activities are even more engaging when children can see the connections to their everyday lives. For example, if children are learning about photosynthesis, have them observe and study the leaves on a tree, or if they are learning about magnetism, have them experiment with magnets and metal objects.

5. FOSTER CURIOSITY

Encourage children to ask, "Why?" You can model this curiosity by asking questions out loud. Ask questions that you may or may not know the answer to. Ask, "I wonder....?" And then follow it with, "Let's go find out!" You can set up a science center or take the children out on a nature walk to observe and ask "I wonder?" questions.

BACKPACK SCIENCES

SUMMARY

Science is an important subject for all children, and the best method for children to learn and comprehend scientific ideas is through hands-on, interactive, and experiential learning activities. Such exercises have been shown to improve science knowledge and engagement, especially in younger children.

Children need hands-on materials and manipulatives to help them visualize and comprehend scientific ideas. Allowing students to actively interact with the material will help them remember and retain it, leaving a lasting impression on their minds. Additionally, giving children the chance to experiment and investigate with hands-on materials encourages their innate creativity.

Science education must also include interactive, physical exercises. They not only aid in children's comprehension of scientific ideas, but also foster the growth of interpersonal abilities like questioning, dialogue, listening, and cooperation. Another essential element of scientific education is experiential learning. Children are better able to comprehend the value of science in their daily lives, thanks to field trips to museums and wildlife preserves as well as talks from scientific experts.

I hope we can all agree that scientific education must encompass and include more than rote memorization of textbook material. Scientific knowledge and engagement are effectively promoted through hands-on, interactive, and experiential learning activities, which give the topic more significance and lasting memory for children.

As educators, it is our responsibility to give children the tools they need to build their own conceptual frameworks for science and link their personal experiences to scientific ideas. By doing this, we can support children's passion for science and foster their natural curiosity about the world.

THE IMPORTANCE OF USING SCIENCE TO ENGAGE OUR CHILDREN WITH INCLUSIVITY AND CULTURAL SENSITIVITY

Why is this important?

Since science frequently involves complex concepts that may be difficult for young children to understand, teaching science can be especially difficult. However, educators and homeschooling parents can make science more relevant and interesting for kids by using inclusive and culturally sensitive lessons.

According to research, including aspects of children's cultural backgrounds and experiences in science lessons can aid in their understanding and connection to the subject. According to research by Gay from 2010, culturally sensitive instruction can boost students' motivation, engagement, and academic success. Children can better understand the relevance of scientific ideas to their own lives and experiences by including things like stories, songs, and other cultural artifacts in lessons.

According to Villegas & Lucas, culturally sensitive teaching strategies must go beyond merely recognizing and celebrating students' ethnic backgrounds. The writers contend that it is crucial for teachers to assist students in developing the abilities and attitudes necessary to interact with people from different cultures and learn from them in today's world of rapid change and global connectivity.

They offer a framework for culturally responsive teaching that consists of five key practices: affirming students' cultural backgrounds, posing questions about deficit perspectives, utilizing students' cultural backgrounds as learning resources, fostering critical consciousness, and facilitating learning in multicultural settings. According to the authors, these strategies can assist children in becoming effective and enthusiastic learners in the twenty-first century.

This is supported by a recent piece by Ladson-Billings, in which she updates a framework for culturally responsive teaching and revisits earlier work on culturally relevant pedagogy. She contends that it is more crucial than ever for teachers to be able to connect with and assist their students in today's fast-changing, complex world. Three key components make up her revised framework: cultural competence (understanding and valuing students' cultural backgrounds), critical consciousness (assisting students in developing a deeper understanding of power and oppression), and caring (developing close bonds with students and fostering a supportive learning environment). This study highlights how crucial culturally competent teaching strategies remain in fostering an inclusive and fair learning atmosphere for all students.

Educators can create a more welcoming and inclusive learning atmosphere where all students feel valued and supported by utilizing culturally responsive teaching techniques.

Studies have stressed the significance of training teachers to adapt their teaching strategies to different cultural contexts. Schools can make learning more inclusive and equitable for all students by giving teachers the resources and training they need to build these skills.

It is crucial to use inclusive and culturally sensitive lessons that acknowledge and celebrate the variety of children's backgrounds when instructing science to elementary school children. By doing this, educators can develop a more stimulating and encouraging learning atmosphere that encourages a love of science and enables children to see the application of scientific ideas to their everyday lives. These techniques can significantly impact how your pupils interact with and comprehend the material, whether you are a teacher or a parent homeschooling your children.

Dr. Maria Montessori developed the Montessori educational approach, which places a strong emphasis on individualized, child-centered learning and the creation of environments that foster a child's natural growth. The Montessori method offers a flexible structure that can be adjusted to suit the needs of specific children, making it a great framework for including inclusivity and culturally responsive practices.

In a 2019 research article, Rosenberg and O'Callaghan reviewed the significant role that Montessori education plays in fostering open-mindedness and tolerance of other cultures. The authors' results demonstrate that the Montessori method is based on respect for diversity and encourages educators to work in partnership with students, families, and communities. A supportive, inclusive, and conducive learning environment is often created by Montessori educators by making sure that the classroom is respectful of all cultures and backgrounds. In a different study, Rath (2019) observed how the Montessori method of teaching guides the development of cultural competence. Lessons place a strong emphasis on self-directed learning and develop a sense of curiosity and respect for the world and its many cultures.

Hands-on, experiential learning is a big part of the Montessori method, which can help children from different cultures become interested. Built into the Montessori curriculum are opportunities to explore and learn about the world around them. By cultivating a shared sense of wonder and curiosity, Montessori educators assist in bridging cultural divides.

The Montessori method is an ideal place for educators to incorporate inclusivity and culturally responsive practices, as it provides a framework that is rooted in respect for diversity, emphasizes individualized learning, and fosters a love of learning. Naturally intertwined in the curriculum, children develop a sense of curiosity and respect for the world and its many cultures.

My Aha Moment

I was teaching ecosystems in a 4th through 6th grade Montessori class, and I wanted to make sure all of the children could connect with the material. I decided to use a variety of teaching methods, including hands-on activities, games, and group discussions, to help the students understand complex concepts.

We were focusing on the idea of interdependence in an ecosystem-how different organisms rely on each other for survival. One of their follow-up activities was to create their own ecosystems.

Jin was an exchange student who had recently joined the class from Korea. He was initially hesitant to participate feeling like he didn't know enough English to fully engage in the activity. I noticed Jin's hesitation and made an effort to include him in the group and help him participate.

As the groups worked on their ecosystems, I noticed that Jin had a deep understanding of how different organisms interacted with each other in their ecosystem. Jin's ecosystem had an unusual combination of plants and animals that were atypical in Southern California. When I asked Jin to explain his choices, he spoke about Korea and how the animals and plants in that ecosystem were different from what was present in the U.S.

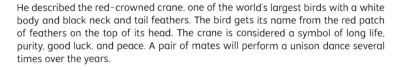

He described the red-crowned crane, one of the world's largest birds with a white body and black neck and tail feathers. The bird gets its name from the red patch of feathers on the top of its head. The crane is considered a symbol of long life, purity, good luck, and peace. A pair of mates will perform a unison dance several times over the years.

He told us about the Siberian tiger, a mesmerizing animal and one of the largest cats in the world. In Korean culture, the tiger is a protective guardian that symbolizes power and strength. Because of the similar shape of the Korean peninsula and the tiger, Korea is known as the land of tigers. Unfortunately, because of poaching and habitat loss, there are no longer any tigers in the wild in the forests of Korea.

It was fascinating to hear about the symbolism of animals and how they reflect the culture. Jin's ecosystem reflected on his own experiences and cultural background, which was an excellent way to share with his classmates.

I seized the opportunity to ask Jin more about his experiences and about Korea. When It was time for Jin to present his ecosystem, because of the supportive learning environment we had created, he felt comfortable enough to include how this ecosystem reflected his cultural background.

As Jin described his ecosystem, the other students in the class were captivated by the various animals and plants they had never heard of before since these animals weren't native to the Sounthern California region. The ecosystem reflected Jin's native country, and he described how each organism depended on the others for survival. Jin was able to confidently reply to the students' questions while feeling proud of his hard work and cultural background.

After this project, we continued including Jin's cultural background in the lessons to better help Jin comprehend the material. The other kids had the opportunity to learn more about Korea and developed a deeper knowledge of both Jin and Korea.

This celebration of ethnic diversity gave Jin more self-assurance and engagement. The other kids were able to relate to the subject matter on a more intimate level and develop a deeper grasp of it.

5 Tips to Engage with Cultural Sensitivity

No matter what their ethnic backgrounds or preferred learning methods are, it is our responsibility as educators to give our students access to a well-rounded education. It is essential that we design science lessons that are relevant and interesting to all of our students while also teaching key scientific ideas.

In order to accomplish this, I have developed 5 suggestions to assist you in designing science lessons that are pertinent, interesting, and less intimidating for all students. By using scientific ideas and examples that are pertinent to the children's real-world experiences, you can include a wide variety of viewpoints and experiences in your science lessons. You can also urge children to bring their own cultural knowledge and experiences into the classroom. We will also go over how to use technology and multimedia to introduce a variety of viewpoints and experiences into the learning environment, as well as how to create an inclusive classroom environment that values and honors diversity.

Join us as we examine these tips and discover how to develop science lessons that inspire and engage all children, whether you are a new teacher just getting started or a seasoned veteran seeking to improve your science lessons. By putting these suggestions into practice, you can establish a learning environment in which each child feels valued, heard, and engaged. I firmly believe that by developing more diverse and culturally sensitive science lessons, we can encourage our students to follow their scientific passions and show them how applicable science is to everyday life. We can all work together to establish a welcoming and encouraging learning environment where every child can succeed.

01 Incorporate a diverse range of perspectives and experiences into science lessons

To demonstrate that science is not just one-dimensional or exclusive to a particular set of people, it is crucial to include a wide variety of perspectives and experiences in science lessons. Children can learn about how science is used and understood by people with various cultural backgrounds, levels of knowledge, and comprehension when educators include a variety of viewpoints and experiences. This can improve the relevance and interest of science to children and foster a more welcoming learning atmosphere in the learning environment. Education professionals can:

Include lessons on how science is applied in various cultural contexts.

Use books, articles, and videos that emphasize varied viewpoints in science.

Use examples of scientists from various cultural backgrounds or underrepresented groups.

Inspire children to discuss their personal scientific encounters and viewpoints.

Example 1

Use cultural artifacts or examples in the science lesson. For instance, you could use examples of cultural artifacts like an African mud cloth, regional pottery, or materials used in traditional clothing from various countries when talking about the properties of matter. This demonstrates the adaptability of scientific ideas to societal settings.

Example 2

When teaching about renewable energy sources, discuss how different cultures around the world have used natural resources such as wind, water, and sun to create energy. Highlight examples from indigenous communities and talk about their relationship with the environment.

Example 3

Invite scientists or experts from diverse backgrounds to speak to children about their work or their experiences in the field. This not only exposes children to diverse perspectives and experiences, but also provides them with real-world examples of how science is used in various contexts.

02 Use science concepts and examples that are relevant to the students' lived experiences

When science concepts and examples are relevant to the children's lives, it helps make science more relatable and interesting for them. It also helps children to see the practical applications of science in their own lives. Educators can do this by:

Looking at examples of science concepts that relate to children's daily lives

Incorporating examples that are culturally relevant or familiar to the children

Encouraging children to share their own experiences and knowledge related to science concepts

Creating projects and assignments that allow children to apply science concepts to their own lives

Example 1

When teaching about simple machines, use examples that are relevant to the children's lives, such as how they use levers to open doors or how inclined planes help them climb stairs. Make sure to bring in concrete examples. A fun activity after the lesson is for children to start making lists of simple machines around the classroom, in their home, or just driving around.

Example 2

When instructing on the human body, go over how the various systems cooperate to maintain health. Use examples of healthy behaviors and customs that are pertinent to their culture and community, such as common treatments or nutritious foods.

Example 3

When talking about the weather, use local examples. Talk about the various weather patterns and how they affect your everyday life, including how the seasons and climatic changes affect your neighborhood.

03 Encourage students to bring their own cultural knowledge and experiences to the classroom

Learning environments are more inclusive and encouraging when children feel that their cultural experiences are acknowledged and valued in the classroom. As mentioned previously, the Korean exchange student, Jin, was able to bring in stories and descriptions of his country's geography and culture.

You can follow this example by:

Encouraging kids to talk about their own science-related experiences and perspectives.

Creating opportunities for children to talk about their cultural experiences and knowledge with the class.

Developing projects and assignments that let children explore science concepts in the context of their own cultural backgrounds.

Giving the class access to resources and materials that represent the diversity of the class's cultures.

Example 1

Start by giving an example of how you apply scientific concepts into your daily life. Encourage the children to share their stories and the ways that their ethnic upbringing has shaped their beliefs and practices.

If the topic of study is renewable energy sources, such as solar power and wind power being used to power data centers and reduce carbon footprint, a Nigerian child can talk about how their family or family members use solar-powered lamps to light up their homes in areas with limited or no access to electricity. They can share their experiences of how they use of renewable energy in Nigerian communities and the benefits it brings to their daily lives.

The child or class can then further explore how renewable energy is used in Nigeria. Children can learn more about various renewable energy sources—including solar power, wind, hydropower, and biomass—and how other countries use them.

Provide children with the opportunity to learn about and share a scientific advancement or finding that has its roots in their culture. Encourage them to learn more about their past, their culture, and the ways their culture has influenced science.

If the class is learning about biodiversity, the educator can discuss the importance of how to prioritize using products that are sustainably sourced and unharmful to the environment, such as buying locally grown produce and supporting companies that prioritize environmentally friendly practices.

The educator can encourage children to share their experiences and stories related to biodiversity and ways their ethnic upbringing has shaped their beliefs and practices. For example, a Peruvian student can talk about how their family has traditionally used natural resources from the Amazon rainforest for medicinal purposes. The child (or perhaps a parent as a guest speaker) can share knowledge of the various plants and herbs that are used, and how they are gathered and prepared for medicinal use.

This allows children the opportunity to learn more about the unique ecosystems found in Peru, such as the Andean cloud forests and the Amazon rainforest where various plant and animal species live. Another example is the scientific discovery of the properties and benefits of quinoa. Quinoa is a grain that is native to the Andean region of South America, including Peru. It has been a staple food source for Andean people for centuries and has recently gained worldwide popularity for its nutritional value and versatility in cooking.

Example 3

Have children study and present scientific subjects from various cultural vantage points while working in groups. Encourage them to talk about what they have learned and how it pertains to their personal experiences with the class.

One time, I told the class about how I find my own spiritual peace hiking and exploring natural landscapes. Understanding geology helps me appreciate the features of the landscapes. For example, various volcanic and geothermal features are found in many parts of the world, and these create wonderful places to visit, such as geysers, hot springs, and lava fields. This creates a place for energy production, tourism, and cultural significance.

Children can then start to share or investigate various geological features, history and formations found in different parts of the world. Children can include personal or researched information on how the unique geology of the various countries influences culture, society and daily life. For example, volcanos and hot springs are found in New Zealand, Russia, Iceland, Japan, and the United States. In Iceland, the famous geothermal spa Blue Lagoon, is a popular tourist attraction used for relaxation and health benefits. Children can analyze the environmental and economic impact of its use.

Foster an inclusive classroom environment that values and celebrates diversity

04

An inclusive classroom environment is one where all children feel welcomed, valued, and supported. Educators can create this kind of environment by:

Encouraging students to respect and appreciate each other's differences, establishing clear expectations for respectful and inclusive behavior

Incorporating materials and resources that reflect the diversity of the class

Providing opportunities for children to learn about and celebrate different cultures and traditions

Addressing instances of discrimination or bias in the classroom

Example 1

Display posters and visual aids that represent diversity in the classroom, or even better, have the children create them. Celebrate cultural holidays and events by including everyone in classroom decorations and activities. For example, when studying sustainable agriculture, it is important to protect the enviroment and ensure food security. This includes farming practices that are environmentally responsible, economically viable, and socially just.

To foster an inclusive learning environment that values and celebrates diversity, children can create posters and visual aids that represent the different sustainable agriculture practices used in different countries—for example, India. Research topics include crop rotation, organic farming, and water conservation.

Through this activity, children will appreciate and celebrate diversity in terms of sustainable agriculture practices and human society. It's important to recognize and learn to value different cultural practices and traditions.

BACKPACK SCIENCES

Example
2

Use inclusive language and avoid stereotypes when teaching science concepts. Acknowledge and celebrate diversity by inviting guest speakers from diverse backgrounds to share their experiences. For example, geology is the study of the earth's physical structure, processes, and history, including rocks, minerals, and landforms that make up the planet. When teaching geology, educators can use inclusive language that avoids stereotypes and promotes diversity. For example, instead of referring to the earth as "Mother Nature," which reinforces gender stereotypes, educators can use gender-neutral language such as "the natural world" or "the earth's systems."

Educators should acknowledge contributions of scientists from diverse and underrepresented backgrounds. For example, in the early 20th century, Annie Jump Cannon was a pioneering astronomer who manually classified more stars in a lifetime than anyone else with a total of over 350,000 stars. Her work and dedication helped women gain acceptance and respect in the scientific community, paving a path for future female astronomers.

Example
3

Implement group activities that promote teamwork and collaboration, regardless of cultural backgrounds or abilities. Encourage children to work together and learn from one another while also respecting each other's unique qualities.

When studying engineering concepts, such as mechanics, energy, and motion, have children design a Rube Goldberg machine. In groups, children can design and build a complex device that performs a simple task in a multi-step, convoluted way. Children will need to brainstorm creative solutions. By working together, students draw on other's strengths and ideas while promoting collaboration and critical thinking skills. This creates an inclusive learning environment where all children feel valued and respected.

05 Use technology and multimedia to bring a diverse range of perspectives and experiences into the classroom

Technology and multimedia can be used to help children explore diverse perspectives and experiences in science. (Please take note that this should be done with adult supervision. All websites should be reviewed by an adult.)

This can be done by:

Using videos, podcasts, and other multimedia resources that highlight diverse perspectives in science. Different cultures and countries may address scientific concepts differently.

Incorporating interactive activities and simulations that allow children to explore science concepts in different cultural contexts. Understanding your students' culture, language and experiences will help you.

Encouraging children to use technology to explore and share their own cultural knowledge and experiences related to science. This may include researching and presenting their follow-up activity.

Providing access to online resources that reflect the cultural diversity of the class.

Example 1

Use video clips and documentaries to show diverse perspectives on scientific topics. Provide discussion questions that encourage children to reflect on what they have learned and how it relates to their own experiences.

Example 2

Incorporate digital tools and interactive games that help children learn scientific concepts in a fun and engaging way. Use tools that are accessible to all students, regardless of their background or ability.

Example 3

Use social media platforms and online forums to connect with other classrooms and scientific experts around the world. Encourage children to share their perspectives and collaborate with others to learn and explore different scientific concepts.

BACKPACK SCIENCES

COMMON CHALLENGES IN INCORPORATING CULTURAL DIVERSITY

I DON'T HAVE ANY RESOURCES OR MATERIALS.

One common challenge educators may face is a lack of resources or materials. They'd like to support the integration of diverse perspectives and experiences in their science lessons, but this may be difficult due to a limited budget, a lack of access to relevant materials, or limited time to develop and prepare new materials.

Solution:

Online lesson plans, videos, or articles may be an option for free or low-cost resources. Backpack Sciences (www.backpacksciences.com) has a full library of ready-to-go lesson plans. Look for materials that showcase diverse perspectives and experiences in science. Discuss the issue with your colleagues and homeschool community. Consider collaboration with other teachers in your school or community, or reach out to community organizations for support and resources. Parents in the classroom or other parents in the homeschooling community may have materials you can use. Additionally, you can consider adapting existing materials to make them more inclusive, such as by modifying images or examples to be more culturally relevant. This may also be an eye-opening experience for the children—they can help with altering the materials.

I'M GETTING SOME RESISTANCE FROM STUDENTS OR PARENTS.

A challenge that an educator might run into is pushback from students or parents who aren't used to seeing different points of view and experiences brought into science lessons. This resistance can come in the form of disinterest, skepticism, or even pushback against the inclusion of certain themes or points of view.

Solution:

The Great Lessons created by Dr. Montessori are considered a foundation for the Montessori curriculum each year. These five "impressionist" stories are springboards presented at the beginning of each year, and the yearlong curricula for science, math, and language are based upon them. They are given each year for all children ages nine to twelve years old. One year, I had an adult learner, Maria, earning her Montessori teaching credentials. She faced resistance with her head of school and parents of the children in the classroom. Science in a Montessori classroom is typically taught from a secular viewpoint. The First Great Lesson describes one idea of how the universe was created.

Educators can solve this problem by talking to children and their parents openly and honestly about how important and valuable it is to have science lessons that are inclusive and sensitive to different cultures. This is a wonderful topic to address in general at Back to School Night—or on a more detailed Parent Education evening, depending on the community you teach at. Show how these lessons can help all children, no matter where they come from. Our goal as educators is to encourage children to be interested and excited about learning. Invite the family to bring in something of their culture. This can help build trust and a sense of belonging, which can help you get past resistance.

Maria was able to emphasize and confirm that in addition to the traditional First Great Lesson, she introduced several different creation stories from different cultures and countries - for example, for example, the biblical and Native American creation stories. The parents and head of school saw that Maria was culturally competent and inclusive; she was not teaching from only one cultural perspective.

I DON'T HAVE EXPERIENCE OR HAVE VERY LIMITED TRAINING.

A third challenge is that educators may feel they have a lack of training or experience in creating and implementing inclusive and culturally responsive science lessons. This may be due to a lack of professional development opportunities or simply a lack of confidence in their ability to create and implement such lessons.

Solution:

One solution is to look for professional development opportunities that focus on diversity and inclusion in science education. Connect with other educators or organizations that focus on inclusion. Experienced colleagues may have advice on workshops or resources.

BACKPACK SCIENCES

5 QUICK FIXES

Five small, doable suggestions that can lead to quick results for educators or homeschooling parents

1. START SMALL

Choose one science concept or topic and find a way to connect it to the children's lived experiences or cultural backgrounds. This can be done by simply asking a child about their experiences or by doing a quick Google search to find examples related to their culture.

2. MIX UP YOUR EXAMPLES AND VISUALS

Be intentional about incorporating a diverse range of examples and visuals in your science lessons. This can include pictures and videos that feature people of different ethnicities, cultures, and genders. You can also use examples from different countries and cultures to illustrate science concepts.

3. MAKE CULTURAL CONNECTIONS

Encourage children to share their cultural knowledge and experiences related to science topics. For example, if you are teaching about the water cycle, you could ask children to share how water is viewed and used in their culture.

4. PRIORITIZE STUDENT-CENTERED LEARNING

Student-centered learning is an approach in which the child chooses what to study based on why that topic might be of interest and how they'd like to explore it further. Give children opportunities throughout the day to share their experiences and perspectives. Invite children to give input and share their experiences and thoughts in a nondisruptive way—perhaps by raising their hands during breaks in the lesson. Reflections and thoughts might be shared during the introductory lesson or just before leaving for the day.

5. REFLECT AND REVISE

Take time to reflect on your lessons and how they incorporate inclusive and culturally responsive practices. Ask for feedback from your students and be open to making revisions based on their input. Continuous reflection and revision can help you grow as a culturally responsive educator.

SUMMARY

According to research, children should be taught science in a way that is engaging and pertinent to them using culturally considerate teaching methods. This approach aims to create a welcoming and empowering learning environment using each student's distinct cultural heritage.

Including a range of viewpoints and experiences in science classes is one way to accomplish this. It is important to teach science concepts in a manner that is applicable to the experiences that kids will have in the real world. We want them to understand how science is relevant to their daily existence. Students feel appreciated and recognized when we encourage them to share their cultural experiences and knowledge in the classroom.

The Montessori philosophy values inclusivity and variation. In a Montessori education, which stresses a child-centered approach to learning, children are given the freedom to choose how they want to learn. According to this way of thinking, every child is unique and deserves praise for their remarkable talents. Additionally, because it increases students' involvement, hands-on learning is crucial to a Montessori education.

The creation of a welcoming classroom atmosphere that values and celebrates diversity is a key component of culturally responsive teaching. This involves providing a safe and welcoming learning environment where every child feels valued and respected. By encouraging children to engage favorably with one another and by setting an example of inclusive attitudes and behaviors, educators can foster this type of environment. The development of a feeling of community is aided by encouraging kids to see themselves as a part of a larger group.

Although culturally responsive education has many benefits, it can be challenging for teachers to implement. Common barriers include lack of training, money, and time. To overcome these challenges, teachers can start with simple, doable activities, like combining mixed media elements or incorporating various models and points of view into their illustrations. Additionally, they can seek amazing professional development opportunities to examine socially responsive teaching examples.

Finally, implementing culturally sensitive teaching methods is essential for developing a more welcoming and interesting learning atmosphere for children. Science lessons can be exciting, relevant, and important for all children by valuing diversity, celebrating it, and including a variety of viewpoints and experiences. Together, educators can pave the way for a better future for all children.

BACKPACK
SCIENCES

Chapter 03

INTERDISCIPLINARY

A Glimpse Into My World

I work constantly to create lessons that will engage and inspire my students. In one memorable lesson on the Amazon rainforest, I was able to incorporate math, geometry, science, art, writing, reading comprehension, geography, history, spelling, music, movement, and speech. Sounds like a lot of work, right? Well, it's not, and I'll tell you how I did it.

Before the class started, we read a book about the Amazon rainforest and talked about its special geography and ecology. We examined the hardships of getting by in such a dynamic environment, the different layers of the rainforest, and the animals and plants that call it home. We used geometry to model the various rainforest layers: the emergent, canopy, understory, and forest floor. We analyzed the shapes, patterns, and spatial relationships found in the arrangement of trees, foliage, and vegetation at each layer.

In the science portion of the lesson, we learned about the different plants and animals that live in the Amazon rainforest, including jaguars, sloths, anacondas, and toucans. We discussed the importance of preserving this ecosystem and the challenges facing it due to deforestation and climate change.

To reinforce this knowledge, we used art to create drawings and collages of different animals and plants that live in the rainforest. Each child was encouraged to choose a plant or animal and draw it with specific details and labels.

Incorporating writing and reading comprehension, we read articles about the culture and history of indigenous people of the Amazon rainforest. We discussed the challenges they face and their important role in preserving the rainforest. For spelling and vocabulary, we created word lists related to the Amazon rainforest.

In the history portion of the lesson, we learned about the history of exploration and colonization of the Amazon rainforest, including the work of naturalists like Charles Darwin and Alexander von Humboldt. We also talked about the different cultures that have inhabited the rainforest over time.

Finally, children applied their speech skills by presenting their findings and collages to the class. Each child prepared a short speech about the plant or animal they chose and why it is important to the rainforest ecosystem. One group of children performed a choreographed dance with Amazonian music.

At the end of the project, the kids were engaged and enthusiastic about the Amazon rainforest. They were well-versed in the geography, ecology, society, and history of this significant ecosystem. Additionally, I noticed great advancements in math, geometry, physics, art, writing, reading comprehension, geography, history, spelling, and speech.

As a teacher, this instruction served as a reminder of the value of interdisciplinary teaching strategies. Combining various topics and assisting kids in seeing links between them can foster critical and creative thinking.

Why is this important?

Cross-curricular activities are also referred to as "interdisciplinary" by educators. Interdisciplinary approaches must be incorporated in order to motivate and engage students in education. Interdisciplinary lessons demonstrate that student comprehension, engagement, and retention of science concepts are enhanced when other subjects are integrated.

Math, art, history, geography, language arts, and other subjects can all be connected to a wide range of scientific ideas. By combining it with other subjects, children can see how science is used in real life and how it affects their daily lives.

Children's engagement has also been shown to improve with cross-curricular strategies. Children are more likely to be interested in and engaged in the learning process when they are exposed to topics that are relevant to their daily lives. Children can better comprehend and appreciate the value of science if they are shown concrete examples of how science affects the environment or is something that they interact with every day—for example, food or their pets.

This type of instructional approach may also lead to better knowledge retention. Children who have numerous opportunities to interact with and practice new material are more likely to retain it, according to Wright et al. (2020) and Kiili et al. (2019).

Numerous studies have examined the efficacy of interdisciplinary methods in education. According to research by Wright et al. (2020), children's engagement with and understanding of mathematical ideas increased when science and engineering practices were incorporated into elementary mathematics instruction (Wright et al., 2020). In a similar study (Vein, Kiili et al., 2019),it was discovered that when science classes included virtual reality simulations, kids were more interested and motivated to learn.

In conclusion, it has been shown that interdisciplinary methods of science education are effective at motivating students and involving them in the learning process. When educators relate science to other subjects, children can better comprehend how scientific ideas can be used in the actual world. This increases their interest and aids in knowledge retention.

Montessori influence

The Montessori philosophy is well-known for its emphasis on interdisciplinary teaching methods and child-centered, hands-on learning. In "cultural albums," which are lesson plans that cover the subject areas (art, history, biology, geography, and physical science), there is a natural weave of cross-curricular concepts and activities. A growing body of peer-reviewed academic research suggests that educators should emphasize this Montessori method when teaching.

In 2006, K. K. Rathunde and M. Csikszentmihalyi examined the effects of a Montessori-based interdisciplinary curriculum on children's learning outcomes. The study found that students who were taught in the Montessori way were academically more advanced than students who were taught in more traditional ways. According to their findings, the Montessori method's emphasis on integrating various subject areas and hands-on, experiential learning may have contributed to the positive outcomes.

Another study by Linnenbrink-Garcia L., Durik, A. M., & Conley, A. M. (2017), published in the Journal of Educational Psychology, looked at the relationship between student motivation and interdisciplinary learning. According to the study, students who received instruction using an interdisciplinary approach exhibited higher levels of intrinsic motivation and greater levels of learning engagment than those who received instruction using more conventional methods. The researchers suggest that the interdisciplinary approach helps to create a more holistic and meaningful learning experience, which can contribute to greater student engagement and motivation. Montessori philosophy teaches interdisciplinary principles by emphasizing the importance of integrating different subject areas in order to create a more holistic and meaningful learning experience. Children should be exposed to a wide range of subject areas, including language, math, science, geography, history, art, and music, in order to develop a well-rounded understanding of the world and see that everything overlaps.

In Montessori learning environments, children are encouraged to explore different subject areas through hands-on activities, rather than being taught in a rote, prescriptive manner. For example, a lesson on the solar system may incorporate aspects of math, science, language, and art, allowing children to learn about the planets and their characteristics, as well as developing their writing and drawing skills.

In addition, Montessori classrooms often feature mixed-age groupings, which allow children to learn from and collaborate with each other. This creates an environment in which children can share their knowledge and learn from each other, fostering a sense of community and a love of learning.

Fortunately, the Montessori approach to education is well-suited to interdisciplinary approaches to teaching, as it emphasizes the importance of hands-on, experiential learning and the integration of different subject areas... but this doesn't mean it's easy. I hope that you can look at this chapter as a guide for you to start structuring your lessons or units of studies.

5 Tips to Make Your Lesson Interdisciplinary

01 Identify natural connections between subject areas

Find connections between various subjects, such as math and science or literature and history, and incorporate these connections into your lessons.

Example 1

Designing a Sustainable Garden

Explain how designing a sustainable garden can link science, math, art, and writing.

A — Start with plant (biology) basics, which include soil types, sunlight, water, and photosynthesis. Calculate garden space and supplies. Students can measure the area, measure the perimeter, and determine garden materials (math).

B — Ask children to make a garden proposal using a budget, materials list, and community benefits (math and communication). This will improve their writing and promote critical thinking about project success (language).

C — Invite children to record plant height, demands on water, weather, and other environmental aspects as they grow (science). To comprehend garden patterns and trends, they can create graphs, tables, and other visual aids (math).

D — Document plant information, including height, drawings, health, etc. in scientific notebooks. Earthworms, insects, birds, and rabbits may visit the garden. Encourage students to watch and record the species' behavior, including movements and reaction to surroundings. This will teach them about variety and interdependence (interconnection in sciences, data collection and analysis).

Example

2

The Science of Cooking

Identify the natural connections between chemistry, math, and language arts by exploring the science behind cooking.

A — Start by discussing the science behind cooking, including how ingredients react to heat and other factors, such as other ingredients.

B — Children will measure ingredients calculate cooking times (math).. What if they need to double or halve a recipe? (math)

C — Invite children to write about their cooking experiences, including the science behind their favorite recipes, what worked well, and what didn't (language). They can also write instructions for making their favorite dish or create a recipe book with their classmates (language, art/graphic design).

D — Have the children research a famous chef and write a biography. They can also explore the cultural and historical significance of different cuisines and dishes (research, social studies).

Example 3

Architecture

A — Introduce the topic of architecture and its impact on society and culture. Share relevant information, such as how buildings are designed to serve different purposes and the historical context of famous buildings and architects.

B — Discuss how different subject areas can be used to approach the science of architecture, such as how buildings are designed to withstand different forces (science), identifying and calculating angles and measurements for different architectural designs (math), exploring the shapes and forms used in architecture (geometry), documenting and analyzing the design of buildings (photography),

C — Invite children to work in groups to research and develop their own architectural designs, incorporating the knowledge and skills from the different subject areas. For example, they might design a community center, a museum, or a residential building. They should explain the scientific and mathematical principles behind their designs and present their work to the class.

D — Organize a field trip to a location where there is unique architecture. Suggest that the children to take photos and document the different shapes and forms used in the architecture. They can also explore the history and cultural context of the location. Back in the classroom, propose that they use their photos to point out the geometric shapes and principles used in the architecture.

02 Focus on the child's learning environment and immediate interactions

Children can learn how to have a positive influence on the world around them by concentrating on their immediate environment. When children can predict a direct impact, they have the chance to work together to achieve a common objective. Multi-level projects can promote a feeling of community and responsibility while addressing subjects like science, math, and language arts.

Example 1

Upcycling Project

A — Begin the project by explaining the concept of upcycling, which is the process of taking waste materials and transforming them into something new and useful (economics). This can involve reusing or repurposing items that might otherwise be discarded (environmental science).

B — Brainstorm ideas for upcycling projects (critical thinking). Encourage them to think creatively about the materials they might use and consider the practical applications of their ideas (innovation).

C — Once children have identified potential projects, encourage them to conduct research (technology) to learn more about the materials they will be using, their properties (science), and any potential safety concerns (health).

D — After completing their research, children should develop detailed plans (design and engineering) for their upcycling projects, including sketches and diagrams of their ideas (art).

E — With their plans in hand, children can begin constructing their upcycled creations (engineering). Depending on the scope of the project, this may involve using hand tools, power tools, and other constructing their upcycled creations (engineering).

F — Once their projects are complete, children can test them to ensure they function as intended. This may involve using instruments to measure things like weight, durability, and usability (science and math).

G — Finally, invite the children to present their upcycling projects to the class (presentation skills) This can be an opportunity for them to explain their thought process, describe the challenges they faced, and showcase their final product (language arts).

Analyzing Data in Sports

Example 2

Introduce the concept of analyzing data in sports and explain how it can help athletes and teams improve their performance.

A — Invite the children to select their favorite sport. This may be one they play on the playground, play during physical education class, or watch on TV. Invite the children to collect and analyze data from sports games or events, using statistics and math skills to determine the effectiveness of different strategies and plays. They can also use language arts skills to write persuasive arguments for why a particular strategy is the most effective. This activity can incorporate physical education, math, and language arts skills.

B — Gather necessary materials such as a stopwatch, scorecards, and pencils.

C — Encourage children to collect data (observation) from sports games or events, such as the time it takes to complete a race or the number of successful shots made during a basketball game. Create a chart or tally sheet that can document the data (math).

D — Have children analyze the data they collected using statistics and math skills, such as calculating averages, percentages, and ratios. They can also use graphs and charts to visually represent their findings (math).

E — Ask the children to use their data to determine the most effective strategies and plays in the sport (math). They can also write persuasive arguments for why a particular strategy is the most effective (language arts).

F — Invite children to share their findings with the class and present their persuasive arguments for the most effective strategy (presentation skills, language arts). Encourage discussion and collaboration among students to compare and contrast their findings (math).

Example
3

Sustainability Project for the Classroom

A — Identify areas of the classroom where sustainable changes can be made. This can include conserving energy, reducing waste, or promoting sustainable behaviors (science).

B — Invite the children to research the environmental impacts of their proposed changes and the technology available to support their efforts. They can also explore the ways in which these changes could benefit the learning environment (science, technology).

C — Calculate the cost savings and environmental benefits of their actions. For example, students can calculate the amount of paper saved by using digital resources or the amount of energy saved by turning off lights when leaving the classroom (math).

D — Brainstorm different language arts activities, such as writing persuasive letters or creating multimedia presentations to encourage local government, companies or organizations to adopt sustainable practices (language arts).

E — Implement their proposed changes and monitor the results. For example, they can track, with just people in the class or their family members, the amount of waste reduced or the energy saved over a period of time (science, math).

F — Reflect on the project and discuss what was learned. Encourage children to share their thoughts and experiences and discuss how they can continue to make sustainable changes in their everyday lives (social studies).

03 Encourage collaboration and communication

When children from different experiences and personalities come together to solve complex, multi-level problems, they bring diverse perspectives and knowledge. This then leads to innovative and creative solutions. Collaboration and communication skills are essential in daily life and in the modern workforce. By working on interdisciplinary science projects, students learn to work in teams, share ideas, and compromise. These skills are invaluable for children to succeed.

One way to do this to create a learning environment that encourages children to work together and communicate their ideas. This can include group projects, classroom discussions, and presentations. In casual discussions, children will naturally see connections with other parts of their lives, what they've learned in the past, or other subject areas.

Example 1

Designing a Mars Colony

A — Introduce the topic of space exploration and the challenges of building a sustainable colony on Mars.

B — Invite the children to divide into groups. Have them take on different roles, such as such as engineers, scientists, and builders.

C — Have children research and create plans for a sustainable Mars colony, incorporating topics such as habitat design, water and food systems, and energy production (geography, ecology, science, math, engineering). Encourage them to consider the resources available on Mars and the impact of living in a different environment.

D — Calculate the resources and materials needed for their colony, as well as the costs of building and sustaining it (math).

E — Explore the physical and environmental challenges of living on Mars, including radiation exposure and the effects of low gravity on the human body (science).

Resources: Mars Colony Design Challenge by NASA (https://www.nasa.gov/audience/foreducators/mars-colony-design-challenge.html) and Mars for Educators by the Planetary Society (https://www.planetary.org/space-policy-and-advocacy/mars-for-educators) can be used as reference materials.

Example
2

Investigating Water Pollution

A — Encourage children to divide into groups and provide them with different sources of water pollution, such as oil spills, agricultural runoff, and plastic waste.

B — Discuss the possible sources and impacts of water pollution, including how it affects the environment and human health (science and health).

C — Encourage children to research the causes and effects of their chosen pollution source, as well as potential solutions to prevent or mitigate its impacts (science). Encourage them to think about the scientific, social, and economic aspects of the issue.

D — Invite children to investigate their own local water source (stream, lake, ocean) and map out possible sources of pollution (geography). Consider arranging water quality tests, and keep records within the school to monitor changes over time (ecology, math).

E — Have students present their findings to the class and discuss how their findings relate to larger environmental issues and what they can do to help protect our water resources (public speaking, presentation skills).

F — Encourage children to take action by writing to local agencies or businesses that may be contributing to water pollution, asking what they are doing to reduce their impact and suggesting possible solutions (language arts). This can be a great way to help children learn how to be engaged and responsible citizens.

Resources: Water Pollution Lesson Plan by Backpack Sciences, and local environmental organizations or agencies that can provide information and support for this project.

BACKPACK
SCIENCES

Example 3

The Science Behind Cooking

A — Introduce the subject of culinary science and go over the chemical and physical alterations that take place during cooking (science).

B — Divide the children into groups and give each one a different ingredient to investigate, including its chemical make-up, nutritional value, and cooking applications (botany, chemistry, health).

C — Encourage children to think differently use substitutions, or develop their own recipes. The materials and quantities used, as well as the procedures followed to prepare the recipe, should all be recorded (science, critical thinking, math).

D — Ask the children to present their recipes to the class and make a cookbook with detailed instructions and pictures of the finished products or the culinary process (speech, art).

E — Talk about the links between science and the culinary arts, specifically how understanding chemistry and physics can improve cooking and help create fresh, inventive dishes.

F — Request a presentation from a local chef or food scientist to the students on the science of cooking and how they apply science to their work.

Resources: Unit on the Science of Cooking by Backpack Sciences, www.backpacksciences.com

04 Modify instructional strategy

Be willing to modify based on interest or to allow further exploration.

Example 1 — Biomes Assignment

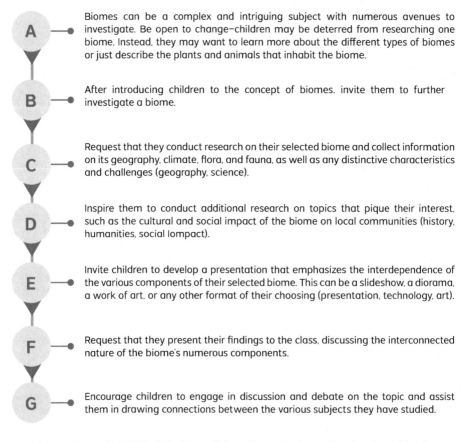

A — Biomes can be a complex and intriguing subject with numerous avenues to investigate. Be open to change—children may be deterred from researching one biome. Instead, they may want to learn more about the different types of biomes or just describe the plants and animals that inhabit the biome.

B — After introducing children to the concept of biomes. invite them to further investigate a biome.

C — Request that they conduct research on their selected biome and collect information on its geography, climate, flora, and fauna, as well as any distinctive characteristics and challenges (geography, science).

D — Inspire them to conduct additional research on topics that pique their interest, such as the cultural and social impact of the biome on local communities (history, humanities, social Iompact).

E — Invite children to develop a presentation that emphasizes the interdependence of the various components of their selected biome. This can be a slideshow, a diorama, a work of art, or any other format of their choosing (presentation, technology, art).

F — Request that they present their findings to the class, discussing the interconnected nature of the biome's numerous components.

G — Encourage children to engage in discussion and debate on the topic and assist them in drawing connections between the various subjects they have studied.

Resources: Biomes of the World by National Geographic (https://www.nationalgeographic.org/encyclopedia/biome/), Biomes for Kids by Science Trek (https://idahoptv.org/sciencetrek/topics/biomes/)

Current Events in Science Education

The educator can take advantage of a news item relating to a science idea that the children are studying to further explore the subject. Here are some actions to take:

A Pick a news item that relates to a science lesson you are presenting. For instance, if there is a news report about a recent wildfire, children can investigate the science behind wildfires, including their causes, how they spread, and their effects on the environment and people.

B Inspire children to learn more and investigate connected subjects. They could, for instance, look into past wildfires in the area and how they have impacted the environment and wildlife.

C Ask the children to debate the connections between current events and the science concepts they are learning as they present their findings and insights to the class.

D Depending on the results, children may have the chance to do something, like volunteer, raise money for relief efforts, or help prevent future occurrences of the same thing.

Example
3

Design Challenge

This challenge involves creating a device or solution to solve a real-world problem. Here are the steps to follow:

A — Brainstorm and identify a problem in the community or in the world that the children are passionate about solving.

B — Invite children to choose one problem they are more drawn to or want to learn more about.

C — Encourage them to research the history and current solutions to the problem and think about how they can improve upon them.

D — Using science and engineering skills, have them design and test their solution, considering factors such as cost, materials, and feasibility.

E — Ask children to present their solution to the class, including a proposal that clearly explains their design and its potential impact (language arts skills).

F — Encourage children to work in teams and collaborate, as well as to give each other feedback throughout the design process.

G — Challenge children to take their solutions beyond the classroom and share them with community leaders, local businesses, or other relevant stakeholders to help make a real-world impact.

05 Identify real-world problems or scenarios that can be approached from multiple subject areas

Look for situations or issues from the actual world that children can investigate using a variety of subject areas. A study on environmental sustainability, for instance, could include science, math, geography, history, and writing components. The following examples allow you to demonstrate the complexity (or AMAZINGNESS!) of how many different topic areas overlap.

Example 1

Food Waste

A — Introduce the issue of food waste and its effects on society, the business, and the environment. Share pertinent data, such as the amount of food wasted annually and its effects, such as food insecurity or a rise in greenhoues gas emissions.

B — Talk about how different subjects can be used to address the problem of food waste— for example, math (calculating the amount of food wasted and its cost), science (examining the environmental effects of food waste decomposition), social studies (examining the cultural and economic factors that contribute to food waste), and language arts (writing persuasive essays on the importance of reducing food waste).

C — Encourage kids to work in groups and create a plan to decrease food waste in their local communities using the knowledge and abilities from various subject areas. For instance, they might figure out how much food is wasted in their community or school, design a composting program, or produce instructional materials to spread knowledge of the problem.

Resource: The Food and Agriculture Organization of the United Nations has a comprehensive report on food waste, which can be found at http://www.fao.org/3/i3347e/i3347e.pdf.

COMPOST

What To Compost

Vegetables	Houseplants	Yard trimmings
Coffee, tea	Fruits	Nut shells
Eggshells	Paper napkins	Paper scraps and cardboard

What Not To Compost

Dairy products	Fats and oils	Eggs, meat or fish bones and scraps
Pet waste	Diseased plants	Produce stickers
Medication	Cigarettes	Broken glass

Example
2

Renewable Energy

A

Introduce the idea of renewable energy and discuss the possible advantages of these sources over conventional fossil fuels. Examples of renewable energy include solar, wind, hydro, and geothermal power. Provide pertinent data on the effects of greenhouse gas pollution and energy use.

B

Talk about how different academic disciplines can be used to study the topic of renewable energy, such as studying the physics and engineering of these sources in science, calculating the effectiveness and cost-effectiveness of various sources in math, examining the political and economic factors influencing the adoption of renewable energy in social studies, and creating persuasive essays on the value of renewable energy (language arts).

C

Encourage children to work in groups to integrate knowledge and abilities from various subject areas into a strategy for implementing renewable energy in their local communities. For instance, they might plan a system of solar panels for their school or community center, research the availability of wind energy nearby, or draft a proposition to encourage the use of renewable energy by nearby companies.

Resource: The National Renewable Energy Laboratory has a wealth of information on renewable energy sources and technologies, which can be found at https://www.nrel.gov/.

Example 3

Public Health

A — Introduce the idea of public health and emphasize its significance for both individual and societal well-being. Talk about contemporary public health issues like disease outbreaks, environmental health risks, and health inequalities.

B — Talk about how various academic disciplines can be used to study public health issues, including the epidemiology of diseases (biology), analysis of statistics on disease incidence and prevalence (math), investigation of the cultural, social, and political influences on public health (social studies), and writing persuasive essays on public health issues (language arts).

C — Encourage children to work in groups to conduct research and create a plan for resolving a public health problem in their local areas, fusing the knowledge and abilities from various subject areas. For instance, they might plan a program to spread vaccination awareness in their school or neighborhood, research local health disparities, or develop a plan to address a local environmental health hazard.

Resource: The Centers for Disease Control and Prevention has a wide range of information on public health issues and initiatives, which can be found at https://www.cdc.gov/

COMMON CHALLENGES EDUCATORS FACE WHEN USING INTERDISCIPLINARY TECHNIQUES

#1

DO I NEED TO HAVE A LOT OF RESOURCES? I DON'T.

Limited resources and lack of funding is one of the biggest obstacles to interdisciplinary science teaching. Schools may not have enough money to give teachers the tools they need to use these methods.

Solution:

Educators can look for grants or area partnerships. Virtual labs and online simulations are also affordable classroom tools. Another choice is to focus on low-resource projects—for example, creating a school garden to explore science, math, and art.

#2

TIME CONSTRAINTS—THERE REALLY ISN'T ENOUGH TIME TO DO THESE LARGE PROJECTS!

Another common challenge that educators face is limited time to plan and implement interdisciplinary approaches in their science lessons. With so many requirements and standards to meet, finding time to develop and implement these lessons can be difficult. You may feel pressure to prioritize content over interdisciplinary learning.

Solution:

One approach is to integrate the content of multiple subject areas into one lesson or project, so that the interdisciplinary approach doesn't take away from science instruction but enhances it. Look for resources that naturally integrate interdisciplinary follow-up activities. Backpack Sciences lesson plans do this (www.backpacksciences.com). Educators can also collaborate with other educators to plan interdisciplinary units that span multiple subject areas, allowing children to explore a topic in depth and make connections across disciplines.

#3

MY STUDENTS DON'T SEEM TO WANT OR KNOW HOW TO DO THESE LARGE PROJECTS. IF THEY'VE BEEN TAUGHT TRADITIONALLY, KIDS MAY OPPOSE INTERDISCIPLINARY SCIENCE.

Solution:

Teachers can engage students by showing them how interdisciplinary learning is relevant to their lives and easy to link. They can also integrate student interests and passions into lessons and projects, helping kids make connections between subjects and apply their knowledge in real life. Educators can also involve kids in interdisciplinary project planning and design, helping them take control and be more engaged in learning.

5 QUICK FIXES

Five small, doable suggestions that can lead to quick results for educators or homeschooling parents looking to make their science lessons interdisciplinary

1. START WITH ONE INTERDISCIPLINARY PROJECT

Instead of revamping your entire program, choose one project to incorporate into your teaching. This is not as overwhelming to you and your students.

2. USE TECHNOLOGY

Many web resources and tools support interdisciplinary learning. Digital maps, data visualization, and modeling software can be used to study geography, math, and science. Take advantage of these resources to enhance your lessons.

3. INTEGRATE MULTIDISCIPLINARY LEARNING WITH PROJECT-BASED LEARNING

Students can work on a project with multiple subject areas instead of single topics. This shows the children how topics relate to one another.

4. PROMOTE STUDENT-LED LEARNING

Let students choose projects or themes that interest them. This can boost motivation and allow them to explore numerous subjects in depth.

5. LOOK FOR CROSS-CURRICULAR/INTERDISCIPLINARY RESOURCES

Backpack Sciences lessons have multiple follow-up activities and tasks that cover multiple subjects. After going over different suggestions and ideas, kids are encouraged to choose one or more and share what they've learned. Visit www.backpacksciences.com for weekly prep-free class lessons.

SUMMARY

In education, an interdisciplinary or cross-curricular method means connecting different academic fields or subject areas in a way that makes sense and fits together. In recent years, this method has become more and more important, especially in the area of science education. Using an interdisciplinary approach helps children get interested in science by making it more relevant to their lives and getting them more excited about it.

Educators can help children see how science is related to other topics—for example, language arts, math, geography, history, geometry, art, and physical movement—by using an interdisciplinary approach. This connection makes science more interesting and less vague for children, and it helps them see how it applies to their everyday lives. When science is taught in a vacuum, it can be hard for children to see how important and useful it is, which can make them lose interest in the subject. But by bringing in other subjects, educators can help children see the bigger picture of how science fits into their learning setting as a whole.

This chapter's "Food Waste" example showed how science can be incorporated with social studies, language arts, and arithmetic. Children can gain knowledge of the scientific procedures involved in composting, the effects of food waste on the environment, and the economics of waste reduction through this endeavor. The project can increase children's interest in science by using a cross-curricular strategy. Children can discover the connections between the topic and other academic disciplines and can see how their science knowledge can be put to use in the real world.

In the "Analyzing Data in Sports" exercise, another example covered in this chapter, children can gather and examine data from sporting events or games. This exercise combines language skills, math, and physical education. Children can develop the ability to use math to evaluate the data and language arts to craft convincing justifications for why a particular tactic is the best one. The exercise is made more relatable and interesting for children by using sports as a teaching tool.

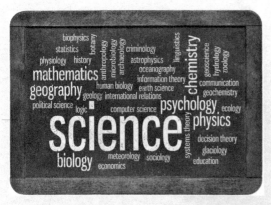

Interdisciplinary science education can make learning more engaging and relevant for children. Children become more eager and interested in learning. They get a more well-rounded education by studying more subjects. Applying scientific concepts to real-world problems helps children build critical thinking and problem-solving skills.

BACKPACK
SCIENCES

Chapter 04

THE POWER OF STORYTELLING WHEN GIVING A LESSON

Why is this crucial?

Many children may find science to be abstract and unrelated to their everyday lives. As educators, we are aware of how crucial it is to involve children in the process of scientific inquiry. A creative instrument for assisting kids in making a personal connection with scientific ideas is storytelling.

We can make scientific concepts more memorable and meaningful for youth by incorporating stories into science instruction. We can inspire awe and excitement about the natural world, igniting a love for science that can last a lifetime.

According to Schmitt, M.C., & Schatz (2018), including storytelling in classes increased students' motivation, engagement, and comprehension. The authors of the study made the case that storytelling makes learning more engaging and meaningful for children by allowing them to make connections between new information and previous knowledge.

In a similar study by Vellutino, A., & Scanlon, D. (2019), the researchers concluded that storytelling can be an effective method for encouraging young children's language development, literacy abilities, and social-emotional development. The authors of the article contend that teachers can develop an atmosphere that stimulates learning and encourages curiosity in children by immersing them in stories.

Incorporating storytelling while introducing scientific concepts can be a powerful way to engage young students and enhance their comprehension of scientific ideas, regardless of your learning environment. Children can be made to see science as a dynamic, exciting subject that is pertinent to their lives and the world around them by fusing narrative, explanation, and exploration.

In order to motivate and engage the upcoming generation of scientists and explorers, let's investigate the power of storytelling in scientific education.

This Is My Why

As a teacher, I was constantly looking for new methods to introduce the wonders of nature to my students. One day, one of my students, Ryan, was especially uninterested when we were studying the life cycle of plants. Ryan sagged in his chair when we began the instruction and resisted taking part. I could tell that he was hesitant to interact with the information.

I made the decision to tell a tale because I knew I had to do something to pique his interest. I started by introducing George Washington Carver, a well-known botanist who was born into slavery and later rose to prominence as one of the most significant scientists of his time. I mentioned Carver's fascination with the natural world and how he devoted his entire existence to studying plants and coming up with new applications for them.

As I incorporated our lesson on plant life cycles into the story of Carver's life, Ryan began to pay closer attention. I discussed how crop rotation, which helped farmers produce crops more effectively and sustainably, was discovered as a result of Carver's research. I described how Carver's research on peanuts resulted in the creation of peanut butter and other goods containing peanuts.

By the tale's conclusion, Ryan had become engrossed. He wished to know more about how plants grow and develop, as well as about Carver's life and work. He began perusing botany books on his own time and was eager to take part in our class activities and experiments.

Ryan's transformation from a reluctant and uninterested student to a motivated and enthusiastic learner was immensely satisfying for me as a teacher. I was able to pique his interest and ignite his enthusiasm for science by using storytelling as a tool to relate our lesson to the real world.

5 Strategies

With the right techniques and tools, you can become a master of science storytelling and inspire your students to develop a lifelong love of science. Below are five strategies for using storytelling to teach science effectively.

Use descriptive language

Using descriptive language is important because it helps children to visualize the story or paint vivid pictures in the child's minds. It also allows children to connect with the content on a more emotional and personal level. We want the child to visualize a scene, "feel" the emotions of how the people felt, and react to any changes or what is happening in the story. This can help to make the information more memorable and can also inspire curiosity and a desire to learn more about the topic.

Tips for educators:

Begin by encouraging the children to use their imaginations and create mental images as you tell the story. **1**

2 Incorporate sensory language to describe sights, sounds, smells, tastes, and feelings to create a more immersive experience for students.

Use vivid adjectives and strong verbs ("as the sun cast a golden glow," "she sauntered into the kitchen,") to describe the action and characters in the story. **3**

4 Use metaphors and similes ("sly as a fox," "quiet as a mouse") to help students understand complex or abstract concepts.

02 Incorporate vocal variety

The power of vocal variety should never be underestimated. Using vocal variety is important because it can help to convey emotion, create suspense, and hold students' attention. A monotonous voice can quickly bore students, while a dynamic voice can captivate their interest and make the story more engaging.

Think back to when you were in any classroom type setting or conference where a speaker is in front of the group, delivering a speech or lesson in a monotonous tone, devoid of any emotion or enthusiasm. It doesn't matter the content-most people's attention will wane, their minds drifting off to distant places. This might be where you start thinking about what you're going to have for lunch.

When the speaker embraces the art of vocal variety, a whole new world opens up. By skillfully employing pitch, volume, and change in voice, the speaker transforms a mundane lecture into an enthralling experience, where people are on the edge of their seats waiting to hear what's next.

Lowering the volume of one's voice can create an air of suspense, drawing the audience closer, eager to hear what comes next. Utilizing different voices and tones for different characters in a story can transport the children to different worlds, igniting their imagination and fostering a deep connection with the content. A sudden increase in volume can emphasize a crucial point, grabbing attention and ensuring the audience grasps its significance.

By skillfully utilizing vocal variety, an educator can transform the learning environment, cultivating a sense of excitement, curiosity, and wonder within the children. It becomes a powerful tool to create a dynamic environment, where you go beyond just facts and a lecture to an experience that touches hearts and minds.

Tips for educators:

Use your voice to add emphasis, convey emotions, and create suspense.

 Use a variety of tones and inflections to add emphasis and interest to different parts of the story.

Use silence strategically to add drama and impact to important moments. This strategy also allows you to think clearly about your story.

03 Practice pacing

Ha, ha-not the pacing describing you walking back and forth! Pacing is the speed at which you're telling your story. You'll want to vary the pace of your storytelling to keep children interested. Pacing is important because it can help to maintain students' interest and attention throughout the story. A story that moves too slowly can be boring, while a story that moves too quickly can be confusing or overwhelming.

Tips for educators:

Begin the story with a slower pace to help students get settled and create an atmosphere of anticipation. **1** Also slow down during important moments or to create tension.

2 Slow down during important or reflective moments to create a sense of gravitas and significance.

Speed up during exciting or action-packed moments to keep students **3** engaged.

04   Use gestures and body language

Using gestures and body language can help to emphasize key points and create a more dynamic and engaging storytelling experience. This can also help students to better understand and remember the information being presented. It can be fun to come up with gestures to go along with scientific vocabulary.

Tips for educators:

Use hand gestures to emphasize important points or to show the size or shape of objects in the story.

Use facial expressions to convey emotion and emphasize the tone of the story. This helps by giving cues to the audience on the emotions of you, the storyteller, or how the other characters feel.

Use body movements to create action and excitement, or to demonstrate processes and movements related to the topic. Even if you're not actually running, swinging your arms as if you were will help the children remember that you were running as fast as you could to get away from the dog that was chasing you.

05   Engage the viewers in the narrative

In the best storytelling, the children are included, and there is some sort of engagement. Children are more likely to recall and comprehend the ideas and the sequence of events when they are actively involved in the storytelling process.

The impressionistic "Great Lessons" are repeated year after year in in Montessori elemetary schools. In the first year, the majority of the kids are amazed. In the first year, the majority of the kids are amazed. By the second year, they might be able to foresee some of the thrilling aspects of the lessons. And by the third year, they might get a little tired of the stories. For this reason, I advise you to involve the older kids in planning the lessons—for example, setting up experiments or demonstrations and taking part in the lesson itself. They may act out scenes or carry props.

Encourage children to ask questions of the teacher or to share their opinions about the tale as another way to engage them. When you use visual aids or props, such as images, diagrams, or tangible items, you boost understanding and comprehension in children. To help me recall the order of the story, I frequently place visual aids in that order. Children can actively engage in the story by acting out scenes, holding up props, showing their own illustrations, or creating their own stories.

COMMON CHALLENGES WHEN INCORPORATING STORYTELLING

WHEN I START TELLING STORIES, I FEEL LIKE I CAN'T MAINTAIN THEIR ATTENTION.

One of the biggest challenges when using storytelling to teach science is keeping children engaged and interested throughout the lesson. This is particularly challenging for younger children or those with shorter attention spans.

Solution:

To maintain children's attention, it's important to use a variety of storytelling techniques, such as incorporating vocal variety, using descriptive language, and practicing pacing. In addition, it can be helpful to incorporate opportunities for children to actively participate in the story, such as by asking questions or making predictions, drawing illustrations, or acting out scenes. Finally, incorporating humor, suspense, or surprise can be effective ways to keep children engaged.

#2 I DON'T FEEL COMFORTABLE WITH TELLING STORIES.

Some teachers may feel uncomfortable or inexperienced with storytelling, particularly if they have not used this technique before. They may feel like they don't possess natural storytelling skills or may lack the ability to captivate the children.

Crafting a compelling story requires creativity, imagination, and the ability to effectively structure and deliver it to hold children's attention. For educators who have not had prior experience or training in storytelling, the prospect of incorporating it into their lessons can be intimidating. Additionally, some educators may feel constrained by time constraints and curriculum demands, perceiving storytelling as an extra component that might disrupt their lesson plans or divert focus from achieving specific learning objectives. The pressure to cover extensive content within limited class time can make teachers hesitant to experiment with storytelling, fearing that it may compromise the pace and completion of the required curriculum.

Solution:

One solution is to start small and practice telling stories in a low-pressure environment, like at home, in front of the mirror, or in front of your family - somewhere you feel comfortable. What has helped me in the past is watching or attending workshops on effective storytelling techniques and seeking mentorship from more experienced educators. To make sure you have the foundation of the scientific concept, you can use existing science-related books or stories as a starting point, and gradually build up confidence and experience over time.

#3 WHEN I START TELLING STORIES, I FEEL LIKE I CAN'T MAINTAIN THEIR ATTENTION.

Another frequent difficulty is how to effectively and meaningfully incorporate storytelling into scientific curricula for kids. If the curriculum is already extensive or there isn't much time to fit in extra tasks, this may be difficult.

Solution:

Using science-related books or tales to introduce new ideas or to review previously covered material are just a few examples of how storytelling can be incorporated into existing science lessons. Additionally, it can be beneficial to work with other educators or look for materials that can offer advice on how to successfully incorporate storytelling into the science curriculum, such as teacher guides or online lesson plans. In order to make sure that the storytelling methods are effective and meaningful for their learning, it can be helpful to solicit feedback from the students. This will allow you to assess their interest and engagement and make any necessary adjustments.

ONCE UPON A TIME

5 QUICK FIXES

Five small, doable suggestions that can lead to quick results for educators or homeschooling parents looking to start using storytelling into their science lessons

 01 Use story prompts related to science concepts to spark the children's imaginations and creativity

Example 1

You might start this with a story of a time you wondered, created, or discussed this.

Imagine you're a scientist exploring a new planet. What would you look for to determine if the planet is habitable?" Children can use their scientific knowledge and imagination to create a story about their journey to the new planet and describe the scientific observations they would make.

Example 2

You might start this with a survival story.

Start with a survival story, for example: "Imagine you're a cell. What would you do to stay alive and healthy?" Children can use their understanding of cell biology to create a story from the perspective of a cell, describing the processes and functions that keep them alive and healthy.

Example 3

Pick out any object and research how it was created. It's always fun to hear how someone's failure, mistake or accident created something we use every day or enjoy.

Imagine you're a famous scientist from history. What did you discover and how did it change the world?" Children can research famous scientists and create a story that highlights their scientific contributions and their impact on the world.

02 Construct engaging science-based story challenges for students

Example 1

Tell a survival tale to get this activity started.

Give children a challenge such as, "A group of astronauts is stranded on a planet with few resources." How are they going to use science to live until help arrives? Working in groups or alone, children can use their scientific knowledge to come up with original answers to the issue.

Example 2

Pick your favorite animal and talk about how it's similar to other animals.

Give a situation, such as "A new species of animal has been found. How can scientists determine the evolutionary background and classification of this new species?" Children can analyze the issue and come up with a remedy using their knowledge of taxonomy and evolutionary biology.

Example 3

Find out the details of a real oil spill and explain what occurred. To children, this may be more relevant if it's closer to where they live.

Give children an issue—for example, "An oil spill has occurred in a nearby body of water. How can engineers and scientists collaborate to clean up the spill and safeguard the environment? Using their scientific knowledge, children can design a plan for cleaning up the spill and reducing its effect on the environment using their scientific knowledge.

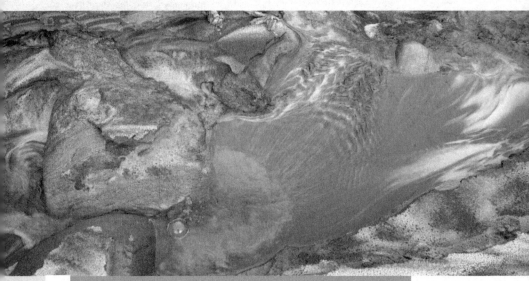

03 Use storytelling to introduce concepts that children may find too abstract

Example 1

To explain photosynthesis to younger children, use a storybook. The narrative can trace a plant's growth and development while emphasizing the importance of water, carbon dioxide, and sunlight.

Example 2

Explain genetics to children using a tale about the finding of DNA. The narrative can outline the scientific discoveries that contributed to our understanding of DNA and how that understanding has been used in contemporary science.

Example 3

Explain infectious diseases to learners using an account of the development of medicine. The narrative can show how ailments—for example, smallpox and tuberculosis—affected people's lives, as well as how discoveries—such as vaccines and antibiotics—altered the path of human history.

04 To promote creativity and teamwork, encourage children to write their own scientific tales and present them to the class

Example 1

Ask classmates to collaborate in small groups to write a narrative about a scientific topic they have studied in class. The stories from each group can be presented to the class, with a focus on the important scientific concepts and an explanation of the creative process.

Example 2

Assign each child to write a brief piece of science fiction that includes a scientific idea they have studied in class. Children can debate how the science was incorporated into the story while sharing their personal experiences with the class.

Example 3

Ask the children to write a comic book or visual novel that explains a scientific idea. Each child can select a different idea and produce a visually appealing narrative that imaginatively and effectively explains the science.

05 Include books and stories about science in the curriculum to improve kids' knowledge of and interest in science

Example 1

Read *The Martian* by Andy Weir to complement a lesson on space travel. Both the theoretical foundations of space travel and the challenges of living on Mars are thoroughly and accurately described in the book.

Example 2

Using a novel, for example, Rebecca Skloot's *The Immortal Life of Henrietta Lacks*, you can teach children about the ethical and philosophical issues that surround medical research. The book tells the story of a woman whose cancer cells were wrongfully used to progress science, and this can spark a discussion about patient rights and informed consent.

Use a text like James Watson's *The Double Helix* to instruct students on the history of the identification of the DNA structure. A discussion about the interactions between collaboration, competition, and ethics in the scientific community can be started using the book's first-person account of a scientific finding.

 # SUMMARY

Children of all ages have long found inspiration and engagement from stories told to them. In order to teach science ideas in a way that is interesting, memorable, and impactful, educators are increasingly using storytelling methods. Educational professionals can help children develop a lifelong love and appreciation for science by using storytelling to bring scientific ideas to life.

One of the major advantages of using storytelling in science instruction is that it makes abstract ideas more relatable and tangible for young learners. Science ideas are more approachable and understandable when they are presented as stories because kids can connect them to experiences and situations they are familiar with. This promotes a greater comprehension of scientific ideas, which can help students retain and apply the information more effectively.

Additionally, storytelling encourages children's interest in science and captures their minds. Children are drawn into the story and develop an emotional attachment to the subject matter when science ideas are presented in a narrative format. This emotional connection can pique children's curiosity and inspire their desire to learn more about science ideas.

Educators can use a variety of strategies to use storytelling in scientific instruction. For instance, they can encourage children's imaginations and ingenuity by using story prompts based on scientific ideas. After they're given a a small prompt or suggestion, children may then be asked to write their own stories, illustrate them, or act out scenes that pertain to scientific ideas.

When educators use stories to explain scientific ideas, they can hold children's interest and make the material more memorable by presenting science ideas in a narrative style. In order to foster a feeling of ownership and engagement in the learning process, they can also urge children to write their own science stories and present them to the class.

Finally, educators can incorporate books and tales about science into the curriculum to increase students' comprehension of and interest in the subject. A more engaging and meaningful learning setting for children can be developed by educators by incorporating books and stories that pertain to scientific ideas.

Although using stories to teach science has many benefits, it can also be challenging for educators. Holding students' attention and understanding how to incorporate storytelling into the curriculum, for example, are two frequent complaints. By utilizing a variety of effective strategies, and seeking out advice and support, educators can overcome these challenges and create a rich and engaging learning atmosphere for their students.

Storytelling is a creative instrument for engaging kids in science and inspiring their love of the subject. By using storytelling techniques, educators can create a rich and memorable learning environment that fosters a profound grasp and respect of scientific ideas. By employing successful methods and strategies, educators can help create a scientifically informed and inquisitive generation.

Chapter 05

FOSTERING ENGAGEMENT IN SCIENCE BY USING CHOICE

As educators, it is our goal to foster in our students a love of learning, enthusiasm, and involvement. Focusing on these qualities is crucial in the field of science education because they canthey can get children ready for the challenges of the future and ignite a passion for the topic that lasts a lifetime. The inclusion of student choice in science education is one strategy that has attracted more notice in recent years.

According to research, giving children options for their educational activities can significantly increase their motivation and engagement. For instance, a study by McComas and Abraham (2004) discovered that students demonstrated greater interest, engagement, and learning gains when they had the opportunity to choose the subjects of their science projects as opposed to students who had their topics assigned. According to findings from Anderson et al. (2015), student choice boosted motivation, autonomy, and self-control in science learning.

Tell me and I'll forget.
Teach me and I'll remember.
Involve me and I'll learn.

- Benjamin Franklin

Including student choice in science education can help children's learning experiences be more personalized while also fostering interest and motivation. According to a 2017 study by Plucker et al., students' interest, engagement, and academic performance increased when they had options in their science education. The authors point out that personalized learning experiences can encourage greater learning and make kids feel more invested in their education.

The development of critical thinking and problem-solving abilities is another advantage of allowing children to choose their own scientific activities. When children are offered options during learning activities, they are compelled to consider their choices carefully and to defend them. Children also develop crucial learning skills—for example, decision-making, problem-solving, and self-evaluation (Lynch and Dembo, 2004).

Incorporating student choice into science instruction can also help children feel more in control and empowered. Children who were offered choices in their learning experiences displayed greater levels of self-efficacy, self-esteem, and academic achievement, according to a 2009 study by Pekrun and Elliot. The researchers point out that encouraging children to feel like they have authority can aid in the development of crucial life skills—for example, self-control and goal-setting.

It is crucial that we as educators take into account how our instructional strategies affect student motivation and involvement. "Choice and agency" is a term used to describe how we perceive and act. These advantages can encourage a lifelong love of learning and better equip our children for the difficulties that lie ahead.

Montessori

The Montessori educational philosophy places a strong emphasis on allowing the child the freedom to investigate and learn about the world around them. A carefully planned environment encourages independence, creativity, and critical thinking. In this setting, choice is essential for fostering engagement and learning, especially when it comes to the instruction of science.

When looking specifically at Montessori science education, it promotes engagement, motivation, and self-directed learning through options. Montessori guides help students feel ownership and power by letting them choose their learning experiences. This can lead to deeper engagement and more meaningful learning.

Using hands-on materials is one way the Montessori method promotes children's choice in science instruction. Numerous materials that appeal to children's senses and promote inquiry and discovery are available in Montessori classrooms. For example, when learning about botany, children may choose to work individually or in small groups with materials-for example, a plant puzzle, a plant labeling exercise, or a plant nomenclature set (Lillard, 2005). These materials are carefully designed to exploration using different senses, self-correction, and other subject areas-for example, handwriting, and Latin root words.

Utilizing open-ended investigations is another way the Montessori method promotes student autonomy in scientific instruction. Children are frequently given open-ended questions or problems by Montessori teachers so they can research and figure out the solutions on their own. For instance, a guide might encourage children to look into the characteristics of water and how they alter as it freezes, melts, or evaporates. Depending on their interests and learning styles, children may experiment, observe, or research these topics (Lillard, 2005). Older students may study soil, plant tropism, and gardening from planting to harvesting and cooking.

Montessori philosophy stresses letting children work at their own pace and choose activities that match their interests and abilities. In science education, this means giving children the chance to pursue their interests and learn more about them. For instance, a dinosaur-obsessed child might read books, build models, or study (Lillard, 2005).

In conclusion, using the Montessori method emphasizes giving children choice. This method encourages children to explore and discover the world in a hands-on, individualized way, promoting engagement, motivation, and self-directed learning. Teachers can foster a lifelong love of learning and prepare students for the future by incorporating Montessori concepts into science education.

Montessori teachers are referred to as "guides" for this reason: we help kids learn. The Montessori method emphasizes hands-on learning and open-ended studies, which foster critical thinking and problem-solving skills needed for science and life.

BACKPACK
SCIENCES

In Her Own Way

I was excited to introduce my class to the topic of genetics. I had planned a variety of engaging activities, including experiments with Punnett squares and discussions about the impact of genetic variations on human health.

But one of my students, a quiet girl named Maya, seemed uninterested in the subject from the start. She sat quietly in the circle with her arms crossed and a frown on her face, barely participating in the class discussions.

I knew I needed to find a way to engage Maya and help her see the relevance of genetics to her life. So, I decided to change up how I presented choices for follow-up activities.

I created a choice board with a variety of genetics-related activities, such as creating a family tree, researching a genetic disorder, and analyzing DNA sequences. I presented the choice board to the class and encouraged everyone to choose the activity that interested them most. This was similar to what I had done in the past, but this time, I used a more visual representation of the choices.

Maya chose to research a genetic disorder, which surprised me. I had assumed she was disinterested in the topic altogether. But as she started to dive into her research, I saw a spark of curiosity in her eyes.

Maya put in long hours over the course of several weeks to complete her assignment. She investigated the effects of cystic fibrosis on people and families at the genetic level. Through a few conversations, I realized that she had encountered cystic disease firsthand. The genetic disease had afflicted her relative.

On presentation day, Maya was a bundle of nerves and anticipation. Her classmates listened intently as she discussed the complexities of genetics and the effects of cystic fibrosis on people and families. She responded to their inquiries with confidence, and I could tell that she was pleased with herself.

Maya was able to delve into her interests and see the relevance of genetics to her life by being given agency over her learning experience. Initially uninterested and tentative, she eventually became passionate and involved. In my position as an educator, that is the best possible result.

10 REASONS
Why choice is important
and how to implement it

01   Personalized learning

It's crucial to support a child's decisions in order to encourage their involvement in science because it provides for personalized learning experiences. By allowing children to choose what they learn and how they learn it, educators can modify their lessons to meet the interests, learning styles, and background of each student. Personalized learning has been demonstrated to increase student motivation, engagement, and achievement, as well as promote a lifelong passion for learning.

Personalized learning refers to an approach where instruction, curriculum, and learning experiences are tailored to meet the individual needs and preferences of each learner. It recognizes that children have unique strengths, weaknesses, interests, and learning styles. Personalized learning aims to provide students with a more tailored educational experience, allowing the child to progress at their own pace, explore topics of interest, and receive support in targeted areas when and where they need it most this style of learning.

3 STRATEGIES TO ENCOURAGE PERSONALIZED LEARNING

Encourage Project-Based Learning

#1

Provide children with a variety of open-ended project options related to a specific science topic or theme. Like Maya, children can choose, the project that aligns with their interests and learning style. For example, if the topic is ecosystems, project options could include creating a diorama of a specific ecosystem, designing and conducting an experiment to study the effects of pollution on an ecosystem, or writing a research paper on a specific endangered species.

A   Introduce the science topic or theme that the children will be exploring.

B   Provide a variety of project options related to the topic or theme. These could include creating a model or diorama, conducting an experiment, designing a research project, or any other open-ended project that allows for student choice and creativity.

C   Explain the expectations for the project, including the timeline, rubric, and any other requirements.

D   Invite children to choose their own project based on their interests and learning style.

E   Provide opportunities for children to check in with you and ask questions throughout the project.

F   Set aside time for children to present their projects to the class and provide feedback to their peers.

Choice Board

#2 Depending on their interests and learning styles, children can choose which activities to complete. For instance, if the topic is the water cycle, activities on the choice board may include creating a visual representation of the water cycle, conducting an experiment involving the water cycle, or composing a poem or song about the water cycle.

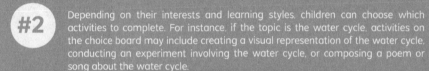

A Introduce the scientific topic or concept that the children will investigate.

B Create a choice board with a variety of relevant activities. Visual representations, experiments, research projects, and creative writing assignments should be included on the choice board to accommodate diverse learning styles and interests.

C Describe the requirements for completing the choice board, including the required number of activities and any other prerequisites.

D Allow children to choose their own activities based on their interests and preferred learning style.

E Schedule time for children to present their completed activities to the class and receive feedback from their classmates.

BACKPACK SCIENCES

Flexible Seating

Give children the opportunity to sit wherever they want during work time, depending on their preferences and needs. While some children might prefer a quieter environment to work alone, others might prefer to remain at a table or on the floor with others. By giving children the choice of where to sit, educators can create a more comfortable and productive learning environment.

A Introduce the idea of flexible sitting and discuss its value in their learning environment.

B Offer a range of seating choices in the classroom, including standing desks, rugs, bean bag chairs, tables, and chairs.

C Allow children to select their daily seats based on their requirements and preferences.

D Encourage children to experiment with various sitting arrangements throughout the year to find what suits them the best.

E Keep an eye on the classroom to make sure children are using the seating choices properly and not interfering with others' learning.

F If they aren't using the seating choices properly, accommodate the child with choices such as a suggested seating arrangement or a timespan for flexible seating.

Gamification

Gamification in a Montessori learning environment can be successful when it aligns with the principles of independent exploration, hands-on learning, and self-motivation. Each child will have the opportunity to make choices, explore their interests, and engage in personalize learning depending of their age and developmental needs.

02 Autonomy

Autonomy in one's own education refers to being able to make choices and take ownership of the learning process. Children can act on their own, without needing help or direction from others. It emphasizes self-direction, self-regulation, and independent decision-making.

In the context of science education, encouraging autonomy means allowing children the freedom to explore and investigate science topics in their own way, using their own creativity and curiosity to guide their learning. When children have autonomy in their learning, they are more likely to be engaged and motivated, as they feel a sense of ownership over their work and can make choices that align with their interests and learning style. Autonomy encourages intrinsic motivation and fosters learning skills such as critical thinking, problem-solving, and self-reflection.

3 EXAMPLES
TO ENCOURAGE AUTONOMY

#1 Inquiry-Based Learning

A Introduce a science topic or concept and allow children to generate their own questions and hypotheses.

B Provide resources and materials for children to conduct their own investigations and experiments.

C Invite children to work individually or in small groups to design and carry out their experiments.

D Encourage children to analyze their data and draw their own conclusions.

E Provide opportunities for children to share their findings with the class and receive feedback.

#2   Child-Led Discussions

A   Introduce a science topic or concept and provide resources for children to explore the topic in depth. Children can also bring in resources—for example, news articles or magazines.

B   Invite children to lead class discussions on the topic, guiding the conversation and asking questions.

C   Encourage children to share their own ideas and perspectives on the topic.

#3   Self-Reflection

A   Provide children opportunities to reflect on their learning and progress in science.

B   Encourage children to set their own goals for their learning and track their progress toward those goals.

C   Provide opportunities for children to self-assess their work and provide feedback to their peers.

D   Invite children to make choices about how they want to improve their learning and what strategies they want to use to achieve their goals.

While personalized learning and autonomy share the common goal of empowering children and promoting individual growth, they approach it from different angles. Personalized learning focuses on adapting the learning experiences to fit the individual, wheras autonomy emphasizes the learner's active involvement in shaping their own experience.

In practice, the concepts can complement each other. Personalized learning strategies can provide children with the flexibility and resources they need to exercise autonomy in their learning environment. By tailoring instruction and learning opportunities to children's interests and needs, personalized learning can foster a sense of ownership and autonomy, allowing learners to take charge of their learning journey.

Relevance

Children's interest in and engagement with science can be affected by how distant it sometimes seems from their daily lives. However, children are more likely to become engaged and motivated learners when offered the chance to connect science with their own lives and interests. Educators can make science relevant and meaningful to children by connecting scientific ideas to everyday situations and letting kids investigate subjects that interest them.

3 EXAMPLES TO
INCREASE RELEVANCE TO CHILDREN

#1 Personalized Research Projects

A Allow children to choose a science topic that interests them and develop a research project to explore the topic in depth.

B Provide resources and guidance as needed but allow children to take ownership of their research.

C Encourage children to make connections between their research and their own lives and interests.

D Provide opportunities for children to share their research with the class and receive feedback.

#2 Science in the News

A Introduce a science topic or concept and ask children to find a news article or story that relates to the topic.

B Invite children to share their findings with the class and discuss how the article or story relates to the science concept.

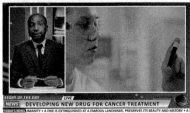

C Encourage children to make connections between the science concept and their own lives and interests.

D Provide opportunities for students to explore the topic further through discussion, research, or project-based learning

 BACKPACK SCIENCES

Science Notebooks

Having a scientific notebook helps children keep science learning relevant by providing them with a dedicated space to document observations, conduct experiments, and reflect on their findings, promoting active engagement, critical thinking, and a sense of ownership in the scientific process.

A Provide children with a science journal or notebook and encourage them to use it to record observations, questions, and reflections related to science topics. Invite them to decorate the cover and make it their own. This notebook will be a space for the children to journal freely, as well as a record of assignments and a documentation of experiments.

B In this particular activity, invite children to choose topics that align with their interests and encourage them to make connections between the science concepts and their own lives.

C Provide opportunities for children to share their journal entries with the class and discuss how their observations and reflections relate to the science concepts.

D Encourage children to use their journal to ask questions and explore topics further through research or project-based learning.

E Provide feedback and guidance as needed to support children in making connections between science concepts and their own lives and interests.

04 Intrinsic Motivation

Intrinsic motivation is the internal drive or desire to engage in an activity solely because it is enjoyable or satisfying in and of itself. Intrinsically driven children are more likely to be engaged, persistent, and creative in their approach to learning. Extrinsic motivation, on the other hand, depends on external factors to encourage behavior and is less effective in promoting long-term engagement and interest in a topic.

To promote child engagement in science, educators have traditionally used extrinsic motivators such as grades or rewards. However, these types of motivators may be ineffective in fostering genuine interest in the topic and may even undermine intrinsic motivation over time. Educators can help to support intrinsic motivation by encouraging child choice and autonomy in science learning, as children become more invested in the learning process and find enjoyment in exploring science ideas on their own terms.

3 EXAMPLES TO FOSTER INTRINSIC MOTIVATION

#1   Science Investigations

A Encourage children to develop a question or problem related to a science concept of interest. Allow them to design and carry out their own investigation to explore the topic.

B Encourage children to develop their own hypotheses and research questions, and provide support and guidance as needed to ensure that their investigations are scientifically sound.

C Invite children to present their findings to the class and facilitate a discussion about their results and what they learned.

D Provide opportunities for children to reflect on their learning and think about how they could apply what they learned in other contexts.

BACKPACK
SCIENCES

#2 Science Debates

A Have children choose from a list of controversial science topics and ask them to research both sides of the issue—for example, the effect of artificial intelligence, organic vs. nonorganic foods, cloning, slowing or reversing aging, or homoeopathic medicine.

B Invite children to choose which side of the debate they want to argue. Provide them with opportunities to practice public speaking and argumentation skills.

C Facilitate a class debate allowing children to present their arguments and engage in respectful dialogue with their peers.

D Encourage children to reflect on what they learned and how their perspective may have changed as a result of the debate.

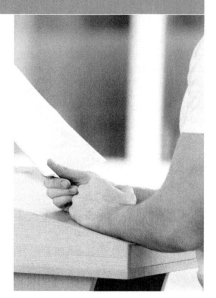

#3 Creative Projects

A Provide children with a broad science concept and allow them to choose a creative medium, such as drawing, painting, or sculpture, to represent their understanding of the topic.

B Encourage children to think creatively and use their chosen medium to explore and express their ideas.

C Provide opportunities for children to share their projects with the class and explain their creative choices and what they learned about the science concept.

D Encourage children to reflect on their learning and how their creative process helped them to better understand the science concept.

05 Critical Thinking

Encouraging critical thinking in children is crucial in today's world, as it enables them to analyze, evaluate, and synthesize information to make informed decisions. By fostering critical thinking skills, children can identify problems and develop creative solutions, leading to increased self-confidence and the ability to make decisions on their own.

However, traditional science assignments often discourage critical thinking by presenting information in a teacher-driven manner, which leaves little room for children to think outside of the box. For example, a science assignment that asks children to memorize facts about the planets after a lecture rather than encouraging them to explore and question the planets' characteristics may not foster critical thinking.

To promote critical thinking in science, what are you implementing to allow the children you work with to explore, question, and investigate?

3 EXAMPLES
TO FOSTER CRITICAL THINKING

#1 "Mystery Bag" Experiment

A Collect a variety of objects, such as a pinecone, seashell, feathers, magnets, or rocks, and place one object in each bag.

B Have children work in pairs or small groups. Give each group a bag and ask them to identify the object using their senses. Record their initial observations and hypotheses about the object.

C The educator can prompt children with questions such as, "What does it feel like? What does it smell like? What does it sound like?"

D Once children have identified the object, ask them to research and present their findings to the class.

#2 "Data Analysis" Challenge

A Introduce the concept of data analysis: the process of examining and interpreting data collected. Explain that data analysis is searching for trends and relationships that will allow the scientist to draw conclusions.

B Provide learners a set of data tables or graphs from either hypothetical studies or real studies from published research articles. Try to select ones that relate to the topic you're currently studying.

C Encourage children to evaluate the data looking at the sample size (how many samples), variability (were there any differences in the sample?), and potential sources of error.

D Have discussions allowing the children to compare and contrast their conclusions and reasoning processes.

E Challenge children to think beyond the presented data and consider additional questions to explore based on the results.

F This challenge is most difficult the first time. After several times, the children will begin to see a pattern on what to do. This is a wonderful activity to begin at the beginning of the year, when basic research, lab and experiment skills are introduced.

#3 Create Your Own Experiment

A Invite students to select a scientific concept or topic they are interested in and brainstorm possible experiments related to the topic.

B Give children the freedom to design and carry out their own experiment on the topic of their choice.

C Provide resources and guidance to guarantee the experiment is scientifically sound. Think about variables, controls, and methods that will make the experiment successful.

D Give children enough time to plan, design, and conduct their experiment and gather data. This also includes analyzing their data, drawing conclusions, and discussing their findings.

E Invite children to present their findings to the class and explain their thought process behind the experiment.

BACKPACK
SCIENCES

06 Collaboration

Science requires creative skills such as collaboration and communication. Group science projects foster an environment where children can cooperate with other children. Collaboration teaches children to communicate, listen, and solve issues. Children can learn more by working together. Science projects also teach children how to communicate plainly and persuasively. These skills are valuable not only in science, but in all aspects of life.

3 IDEAS TO
FOSTER COLLABORATION

 #1 Science Communication Posters

A Invite children to divide into groups of 3-4.

B Provide each group with materials such as poster boards, markers, and images related to scientific topics.

C Invite each group to choose a scientific topic to research and create a poster about.

D Encourage groups to research the topic together, sharing information and ideas.

E Encourage groups to design their posters together, assigning different tasks to each member.

F Set aside time in class for groups to work on their posters and collaborate with each other.

G Once the posters are complete, invite groups to present their findings to the class. This encourages communication and discussion among children.

#2   Science Podcasts

A   Explain what a podcast is and does. A podcast is a digital audio program. It's very similar to a radio show, but you must subscribe to a podcast. You may listen to it anytime you want.

B   Challenge children to create a podcast educating the audience about a science topic of their choice. First, have children research different podcast formats and styles. Provide each group with recording equipment such as microphones and recording software.

C   Invite each group to choose a scientific topic to research and create a podcast about.

D   Encourage groups to research the topic together, sharing information and ideas.

E   Have the group write a script. Have them choose among themselves different parts.

F   Set aside time in class for groups to record their podcasts and collaborate with each other.

G   Once the podcasts are complete, have groups present them to the class, encouraging communication and discussion among students.

H   This may be the start of a school science podcast!

BACKPACK
SCIENCES

#3 Science Peer Review

A Provide or have each group choose a scientific article to review.

B Encourage groups to read and analyze the article together, sharing their thoughts and opinions.

C Have groups provide feedback to the author of the article, discussing what they liked and what could be improved.

D Set aside time in class for groups to collaborate and communicate with each other about their findings.

E Once the peer review is complete, have groups present their feedback to the class, encouraging communication and discussion among students.

07 Critical Thinking

It's crucial to let children approach problems in their own special manner because this can result in more creative solutions and give children a sense of accomplishment. Children are more likely to develop original solutions that might not have been thought of otherwise when given the freedom to think imaginatively and explore their own ideas. Educators can promote a learning culture that celebrates individuality and pushes children to take risks in their learning by allowing for a variety of approaches to problem-solving. Children who learn to believe in their own skills and potential for innovation feel more accomplished and self-sufficient as a result.

Imagine, for instance, that a science class is charged with creating a device to assist in cleaning up a nearby park. One young person might take an engineering-based approach to the issue, designing a device with pulleys and gears that can gather and organize litter. Another child might take a biological approach to the issue and design a device that uses bacteria to decompose and compost organic refuse. These two approaches are both creative and successful, and they each let the child bring something special to the table.

3 PROJECTS
IMPLEMENTING CREATIVITY

Invention Challenge

Ask the children, either individually or in small groups, to come up with a brand-new tool or procedure to solve an everyday issue.

A   Invite children to brainstorm different everyday problems.

B   Encourage them to be original as they brainstorm issues—for example, finding lost items, creating ways to wake up in the morning, or ways to wash dishes without getting your hands wet.

C   Offer assistance and direction as required. Describe the targets, deadlines, and objectives.

D   Allocate specific class time for groups to work together and exchange ideas.

BACKPACK SCIENCES

#2 Science Art Project

Request that children produce a work of art that focuses on a particular scientific idea or theory.

A Encourage the children to use original thinking and a variety of tools and strategies to approach the task in their own special ways.

B Provide a variety of art mediums – paintings, sculptures, or music – to promote diversity and options.

C Provide chances for children to share their work with others, and allot time during class for them to work on their projects.

#3 Solutions to Community Problems

Introduce a problem or conflict that is affecting the neighborhood or the class. Examples include community trash dumps and recycling.

A Encourage the children to come up with new ideas while working in groups.

B Invite the children to share their thoughts.

C The group can decide on any number of the solutions, or even all of them.

D Act and put the concepts into practice.

08 Exploration

It is essential to let children explore various scientific areas and topics. In doing so, we can help them gain a more comprehensive grasp of the subject area. Children are able to better understand how various scientific fields are connected by investigating a range of topics, which can improve their overall knowledge of science. As children begin to view science as a complex and interconnected web of knowledge rather than just a collection of isolated subjects, this can increase engagement and curiosity. Children can also find their own interests and passions in the study of science through exploration, which can inspire a lifelong love of learning.

The educator might, for instance, propose a "science fair" project where students are given the opportunity to investigate and research a range of scientific subjects and present their results to their peers. Any subject that interests a child can be chosen, including environmental science, biology, physics, or chemistry. Children may discover different interests while researching, carrying out experiments, and gathering data throughout the project.

3 PROJECTS
IMPLEMENTING EXPLORATION

#1 Science Exploration Journals

A Introduce the concept of science exploration journals and explain their purpose.

B Provide children with blank journals or notebooks and encourage them to personalize them.

C Set aside time for children to explore different science topics and record their findings in their journals.

D Encourage children to share their journal entries with each other and discuss their findings.

E Provide prompts or questions to guide their exploration, such as "What did you observe?" or "What questions do you still have?"

#2 Science Field Trips

A Present children with a list of science-related locations and their descriptions.

B Have children research background information on the location and what they can expect to learn or see there. Allow children to vote on which location they would like to visit.

C During the field trip, allow children to explore and ask questions.

D Encourage children to take notes or record their observations in a journal or on a worksheet.

E After the field trip, provide time for children to reflect on their experience and share what they learned with the class.

#3 Science Inquiry Projects

A Explain to children the idea of science inquiry projects and what they are intended to achieve.

B Provide children a collection of pertinent scientific questions or let them come up with their own.

C Encourage children to form small groups and present to them the rules and methods they need to perform investigations and experiments.

D Allocate specific class time for groups to work together and exchange ideas.

E Encourage groups to use innovative and interesting presentation methods to share their results, such as a poster, presentation, or model.

F Give children enough time to ask questions and comment on each other's efforts.

09 Confidence

Self-confidence refers to one's self-assurance. It empowers learners to take chances and make decisions. Children are more confident and take control of their science learning when they take charge of their education and shape their learning. They can choose their own path. This confidence can boost self-esteem, science enthusiasm, and interest. Owning their learning makes kids more engaged in it and proud of their accomplishments. When a child designs and conducts their own experiment and presents their results to the class, they may feel proud and accomplished.

Self-confidence can play a role in fostering autonomy, as learners who are confident in their abilities are more likely to take ownership of their learning. However, autonomy and individualized learning go beyond self-confidence. Autonomy encompasses the ability to self-direct, self-regulate, and take responsibility for one's own learning. It requires decision-making, goal-setting, and self-reflection. Personalized learning focuses on tailoring instruction and learning experiences to individual needs and preferences. While self-confidence can facilitate autonomy and personalized learning, they are broader concepts that involve various aspects of the learning process.

3 PROJECTS TO BUILD CONFIDENCE

#1 Choice Boards for Science Projects

A Create a board with a range of science project options—for example, hands-on experiments, research projects, or multimedia presentations. Display it in the classroom.

B Encourage children to select the project that they are most passionate about.

C Offer a project overview document and a rubric outlining the project's requirements.

D Enable children to take charge of their education by letting them establish their own project deadlines and goals. Make sure you evaluate and guide those who might benefit from a little more structure.

E Arrange frequent check-ins with each child to make sure they are progressing and to offer assistance as necessary.

F Allocate time during class to allow learners to work on their tasks and collaborate with their classmates.

#2 Science Interest Surveys

A Give your students a science interest survey to learn more about their interests and motivations in the subject.

B Apply the findings to inform your teaching and give children chances to pursue their interests.

C Enable children to take charge of their education by letting them establish their own objectives and due dates for pursuing their passions. This is a great package for enhancing time management abilities with direction.

D Give children the tools and encouragement they need to learn about and pursue their hobbies.

E Arrange frequent check-ins with each kid to make sure they are progressing and to offer assistance as necessary.

F Allocate time in class for students to pursue their hobbies and cooperate with their peers.

Science Expert Presentations

#3

Children become experts when they do extensive research and become knowledgable in a particular area. This is a wonderful way to boost a child's confidence, as they are now the "expert."

A Explain the definition of an expert. Encourage children to choose a topic of their choice to become an expert in.

B Provide guidelines for conducting research and creating a presentation.

C Encourage children to take ownership of their learning by setting their own goals and timelines for completing their research and creating their presentations.

D Schedule regular check-ins with each child to ensure they are making progress and to provide support if needed.

E Set aside class time for children to work on their presentations and collaborate with their peers.

F Schedule presentation days and provide opportunities for children to ask questions and provide feedback to their peers.

10   Real-World Application

When children have choices in what to learn in science, we encourage them to see the real-world applications of scientific concepts. Science becomes more interesting to children when they see its connection with their lives. This improves understanding and science attitudes. For example, children may study real-world issues such as climate change or pollution. By exploring a topic that interests them and has real-world relevance, children may become more invested in the project and better grasp the science concepts.

3 PROJECTS TO ENCOURAGE REAL-WORLD APPLICATION

#1   Science Project Selection Boards

A   Research science projects with practical uses. The educator can complete this in advance or give children the opportunity to come up with their own suggestions.

B   Make a choice board with a variety of tasks that correspond to the children's interests.

C   Show your students the choice board and go over the practical uses for each project.

D   Let children select the project that most interests them, and allow them time to learn about and consider how their project might be used in the real world.

E   Establish precise expectations and completion dates for the undertaking.

F   Encourage children to share their projects with the class and talk about how their work is used in the actual world.

Science Case Studies

#2

A case study is an in-depth look at one specific study. Researchers can explore real-world applications by delving into the details of a particular situation, providing insights, and understanding deeper the scientific principles.

A Research and identify actual case studies that illustrate how science concepts are applied in various disciplines.

B Show the students the case studies and guide them on how to evaluate and recognize the relevant scientific principles.

C Encourage children to form small groups and give them enough time to evaluate and talk about the case studies.

D Encourage children to share their research with the class and talk about how the science ideas can be applied in everyday life.

#3 Science Service Projects

A Divide children into small groups and have each group choose a real-world issue related to science.

B Provide guidance on how to research and develop a service project to address the chosen issue.

C Allow enough time to collaborate and communicate their ideas with each other.

D Set clear expectations and deadlines for the completion of the service project.

E Set aside class time for children to work on their presentations and collaborate with their peers.

COMMON CHALLENGES EDUCATORS FACE WHEN INCREASING CHOICE FOR CHILDREN

MY STUDENTS SEEM OVERLOADED WHEN I GIVE THEM TOO MANY CHOICES.

Offering too many choices to students can overwhelm them or lead to decision fatigue, making it difficult for them to make a decision or commit to a particular project or activity.

Solution:

One solution is to offer a smaller number of high-quality choices that align with interests and abilities. Make sure all of the available choices are ones that you agree with. This also empowers the children, as they feel like they have some choices and decisions in their education.

I HAVE LIMITED TIME AND HAVE TO STICK TO A SET CURRICULUM.

Curriculum requirements or standards may limit the flexibility of educators to offer choices to students. There is an expected curriculum with topics for each year or semester.

Solution:

Look for ways to incorporate student choice within the curriculum, such as allowing children to choose the topic of a research project or to design an experiment to meet a specific learning objective.

I'VE NEVER DONE THIS WITH MY STUDENTS. THEY DON'T SEEM TO BE INTERESTED.

Some children may resist taking ownership of their learning or may prefer more traditional approaches to science education. This is often because they don't have experience with this type of learning.

Solution:

Educators can provide opportunities for student voice and choice, seek student feedback, and model enthusiasm and excitement for science to help build buy-in and engagement. At the beginning, you'll have to set guideline, schedule check-ins, and give examples. It might seem to be more work at the beginning, but as the children get used to this style of producing their projects, it will become easier with less time for the teacher.

5 QUICK FIXES

Five small, doable suggestions that can lead to quick results for educators or homeschooling parents looking to incorporate more choices into their science lessons

1. ASK OPEN-ENDED QUESTIONS

Try asking open-ended questions that invite multiple potential responses rather than providing a single answer to a question. This allows children to express their own thoughts and fosters critical thinking.

2. GIVE CHILDREN OPTIONS

Let children choose which project to work on or which article to study as part of their learning. This can motivate them more and assist them in taking responsibility for their education.

3. LET CHILDREN ESTABLISH THEIR OWN GOALS

Let children set their own goals for what they want to learn or accomplish. This could help boost drive and give learning more purpose.

4. GIVE FEEDBACK

When children make strides or display growth, give them feedback and encouragement. This may aid in boosting self-confidence and drive.

5. MAKE IT ENTERTAINING

Include entertaining and imaginative activities in your classes, such as interactive games or experiments. As a result, learning may become more engaging and pleasurable.

BACKPACK
SCIENCES

SUMMARY

Allowing more choice in science education has shown to increase child engagement and foster a love of science. This approach allows children to take charge of their education and apply scientific concepts to real life. When children are engaged in their education, they are more likely to participate, ask questions, and take risks, which deepens their comprehension.

Teaching science with more options can help children develop a positive attitude towards science. Giving children a say in what they learn allows them to appreciate it and crave to learn more. This approach can increase curiosity about science and its role in our community.

This method can also bridge academic and practical learning. By applying science in real life, students can see its relevance and importance. This makes the content more engaging and meaningful, improving comprehension and retention.

More options in scientific education might also help in meeting the diverse needs of children. Everyone learns differently, and educators can better accommodate various learning styles and preferences by giving children a wide range of choices to explore. Children may be more involved and engaged as a result, regardless of their individual strengths or challenges.

Additionally, giving children options in their science education can encourage them to participate actively in their education. Children may develop a feeling of ownership and responsibility for their education by being given the freedom to choose the subjects that interest them and to direct the course of their studies. Increased drive, self-directed learning, and improved academic performance can result from this.

Overall, introducing choice into scientific instruction can increase student motivation, engagement, and achievement. It can encourage a passion for learning and heighten a child's understanding of the subject matter, making it simpler for teachers to impart knowledge and for children to learn. Educators can foster a more inclusive and productive learning environment by offering chances for practical application and meeting different learning requirements.

Educators can start by offering children choices to choose from, such as a menu of real-world case studies or a choice board of science projects, to introduce more choices into science education. In addition, educators might let students select the follow-up activity type, such as a written report, a visual presentation, or a practical experiment. By providing resources and encouragement for independent study and exploration, educators can also inspire children to follow their own interests and passions in science.

In conclusion, giving children more options in scientific studies can significantly increase their motivation, engagement, and academic success. Educators can encourage a passion for learning and heighten learner appreciation for the subject matter by giving children the freedom to take charge of their education and offering them chances for real-world application. As a result, a more inclusive and productive learning atmosphere for everyone may be created, making it simpler for educators to instruct and for children to learn.

BACKPACK
SCIENCES

Chapter 06

USING REAL-WORLD ISSUES AND PROBLEMS TO HELP ENGAGE

Why is this important?

In order to motivate and interest elementary-aged children in science education, it has become increasingly important to use real-world issues and problems. As educators work to improve the way science is taught in classrooms, they are increasingly relying on practical, real-world events to help children engage and enjoy science. The significance of incorporating real-world scientific issues and problems into basic science education will be discussed in this chapter.

One of the main advantages of using real-world scientific issues and problems in science education is that it enables children to make connections between the ideas they are learning in the classroom and their surroundings. According to research, children are more likely to develop a deeper grasp of scientific concepts and to recall them in the future if they are given the chance to explore real-world science problems. For instance, in a study conducted by Furtak et al. (2019), elementary students were found to have a better understanding of science concepts and to be more interested in science than students who did not participate in a problem-based science curriculum that focused on real-world issues like climate change and energy in a study conducted by Furtak et al. (2019).

Another benefit of including real-world scientific issues and problems in basic science education is the improvement of kids' critical thinking and problem-solving skills. Working on real-world science problems forces children to think critically and creatively about how to handle issues, which can help them develop the abilities they need to succeed in both science and life.

For example, Gumundsdóttir et al.'s (2018) study discovered that elementary students who took part in a problem-based science curriculum that focused on modern issues like pollution and sustainable energy had better problem-solving skills and were more interested in science than students who did not.

Along with improving understanding and engagement, using real-world scientific issues and problems in the learning environment can support equity and inclusion. By concentrating on issues that are relevant to children's lives and communities, educators can ensure that science education is both accessible and pertinent to all children, regardless of their origin or identity. According to a study by Tyler-Wood et al. (2019), elementary students who took part in a problem-based science curriculum that addressed modern issues like environmental justice and water quality were more engaged in science and felt more empowered than those who did not.

Overall, it has been discovered that including real-world issues and problems in science education has a number of advantages for children, including enhanced comprehension, engagement, critical thinking, and problem-solving abilities as well as equity. As educators continue to work to improve science education, using real-world science problems and issues will become increasingly important in assisting children in developing the skills and knowledge they need to thrive in science and life.

BACKPACK
SCIENCES

From Little Seeds Grow Mighty Trees

One of my favorite long-term units of study is incorporating stream ecology lessons with a service-learning project. I remember vividly the first time I did this. In the classroom, I spent a month concentrating on the value of clean water and how water pollution poses environmental dangers.

At the end of the month, I planned a field excursion for the students to a river nearby so they could witness the consequences of water pollution firsthand. I wanted the children to experience how dumping refuse and chemicals into waterways can damage aquatic life and render water unsafe for human consumption.

One nine-year-old student, Jessica, experienced significant impact from the lesson. Environmental Science had always piqued her interest, but she had never before appreciated the seriousness of the problem. She was motivated to change things as she observed the contaminated water.

Jessica started her own independent study on water pollution over the following few weeks. She read books and papers about how water pollution affects both human health and aquatic life. After learning what she did, she was horrified and inspired to take action to safeguard our water supplies.

Jessica began to discuss water pollution with her family and friends. She urged them to use less water and to be careful with their water use. Her mother laughed as she told me that Jessica yelled at her for taking lengthy showers.

I was delighted to see how the other students responded to Jessica's leadership role and passion. She arranged an all-school event to help clean the local river by removing trash and debris. I was pleased to see my students making environmental protection decisions.

Jessica's enthusiasm for the environment continued for the rest of the year. She began working as a volunteer for neighborhood environmental groups and became a fervent supporter of clean water and ecosystems.

This example served as a reminder to me of the ability of education to bring about positive change in the world. I was happy to see my students making environmental protection a priority. This motivates me to keep involving youth in science by addressing real-world problems. Jessica and her class were successful in making a change to those around them.

Montessori Influence

As it places a strong focus on practical, experiential learning, Dr. Montessori's teaching approach is perfect for integrating real-world practices into science education.

Lillard and Else-Quest (2006) conducted a study on the results of a Montessori education and discovered that these children had higher levels of creativity and performed better on math and science standardized exams. The authors contend that the inclusion of real-world activities and the Montessori method's focus on hands-on exploration and problem-solving support young students' interest and success in STEM fields.

Hohmann and Weikart (1995) conducted yet another investigation into the function of real-world tasks in Montessori education. The authors point out that a Montessori education gives children the chance to participate in practical activities like building, gardening, and cooking, all of which support children's comprehension of science ideas and the value of science in their everyday lives.

Additionally, the Montessori approach emphasizes individualized teaching and encourages children to follow their own interests and passions. Educators can integrate real-world practices, especially if they can tailor the experiences and activities to each child's unique interests and requirements.

The Montessori educational philosophy supports and emphasizes the importance of including real-world uses in scientific education overall. The emphasis on hands-on exploration, problem-solving, and individualized learning provides a strong foundation for integrating real-world actions and experiences into science classes. The Montessori approach also emphasizes the value of fostering young children's interest in and participation in science, which can be accomplished by employing hands-on activities and applications.

5 Tips to Involve Children in Real-World Issues

Education in science is crucial to contemporary culture. Children must therefore be given the information and abilities necessary to succeed in a world that depends more and more on scientific advancements. Many kids, however, find it difficult to relate to science notions because they don't see how these ideas apply to their everyday lives. One method to make science education more meaningful, pertinent, and inspiring is to involve kids in real-world issues.

Children are more apt to be interested in the topic and motivated to learn when they can relate science concepts to everyday situations. Children can benefit from this as they grow in their ability to think critically, solve problems, and work collaboratively in both their scholastic and professional lives. We've gathered 5 classroom-friendly suggestions to assist teachers in getting their students involved in real-world issues.

BACKPACK SCIENCES

01 Identify real-world problems that are relevant to children's lives and interests

When science is relatable and meaningful, children are more likely to be motivated and invested in their learning. One effective strategy for doing this is by identifying real-world problems that children can investigate, explore, and potentially solve.

Community or Personal Surveys

Example 1

Use surveys or a questionnaire to find out what real-world problems affect children's lives. Questions may start off with if there are any health or environmental concerns in their communities—for example, local air or water pollution.

Example 2

News and Current Events

Use current events to expose children to contemporary issues that they can relate to. For instance, you could use a climate change news story to start a lecture on environmental science. Find news and current events on websites like Newsela, The New York Times Learning Network, and CNN 10 as well as in local, regional, and international news sources.

Implement project-based learning (PBL) as a strategy for identifying and resolving real-world issues, as shown in PBL is a child-centered method that includes kids in researching, working together, and applying critical thinking to real-world issues. Projects that focus on local, regional, or global problems can be created by educators. Websites like Edutopia, TeachThought, and Buck Institute for Education all have project ideas and tools.

Sources to Locate National, Regional, and Global Issues

Sustainable Development Goals (SDGs) of the United Nations

01

The UN has established 17 SDGs that must be accomplished by 2030. Global issues, such as poverty, hunger, education, gender equality, and climate change, are all addressed by these objectives. Educators can use the SDGs as a framework to identify local, regional, and global problems that are important to their students. Resources for educators, including class plans and activities, are available on the UN website.

National Geographic Education

02

This website offers free teaching materials for teachers to use in their classes. Lesson plans, activities, and multimedia tools are available for educators on a range of subjects, such as environmental science, geography, and cultural studies. A part of the website is devoted to news and current events.

Global Oneness Project

03

The Global Oneness Project offers free multicultural tales and movies that seek to advance sensitivity and understanding across cultural divides. Teachers can use these resources to get their students interested in global issues that impact people all over the globe. Lesson plans and conversation starters for use in the classroom are also available on the website.

02 Use real-world data and case studies to help children understand the impact of science on their daily lives

By using real-world data and case studies, educators can help children comprehend how science affects their everyday lives and inspire them to take action to solve real-world problems. These global resources can serve as a jumping-off place for educators to find current and relevant data on subjects that matter to their students.

Air Quality

Educators can use current information on air quality to inform students about how air pollution affects both the ecosystem and human health. Governments and organizations all over the world use the Air Quality Index (AQI) as a tool to assess air quality and offer advice on how to remain healthy in polluted areas. Through websites like AirNow (US), AQICN (international), or PurpleAir, educators can obtain real-time air quality information for their neighborhood. Children can better understand the effects of air pollution on their daily lives and be inspired to take action to improve air quality by analyzing this data with their teachers.

Global Warming

Educators can use case studies and data from organizations like the Intergovernmental Panel on Climate Change (IPCC) to teach children about the causes and consequences of global warming. The IPCC is an international organization that brings together scientists from around the world to assess the science related to climate change. The organization publishes reports that summarize the latest research on specific topics—for example, the physical science of climate change, the impacts of climate change on ecosystems and societies, and options for mitigating and adapting to climate change. Educators can use these reports and other data sources to help children understand the global impact of climate change and encourage them to take action to reduce their carbon footprint.

Water Quality

Educators can use actual statistics to show children how water quality affects both the environment and their health. The World Health Organization (WHO) offers statistics and details on the condition of water around the globe, including the standard of drinking water and the effects of water pollution on human health. This information can be used by educators to teach children the value of clean water and how it affects their everyday lives. Children can better comprehend the effects of human behavior on the environment by using the information and resources that are provided by organizations, such as the Ocean Conservancy, on the effects of marine debris and plastic pollution on ocean health.

03 Incorporate storytelling into science lessons

Since stories have historically been used to spread information, morals, and ideas, they can be an effective way to engage children and get them interested in science. Educators and homeschooling parents can help children connect with science ideas on a deeper level and see the practical application of scientific research by incorporating storytelling into science lessons. Children may be inspired to seek further education and careers in science.

Feature Scientists

Use examples of scientists and their efforts from the real world. For instance, educators can discuss Marie Curie and how her discoveries of the radioactive elements polonium and radium changed the course of science and medicine. Learning about specific scientists may motivate children to learn more about science and the ways it can alter the world.

Educators can browse websites like The Story Collider, which features true, individual tales about science, to find tools for incorporating storytelling in science education.

Additional tools on using storytelling in science teaching—including articles, books, and webinars—are available from the National Service Teaching Association (NSTA).

Using Case Studies

Case studies are examples of how scientific theories and ideas have been utilized to handle or solve issues in the real world.

An educator might use a case study to demonstrate the successful elimination of smallpox, a fatal and contagious disease that once threatened millions of people worldwide. The case study looks into the evolution of the disease, the scientific breakthroughs that led to the discovery of a vaccine, and the global campaigns undertaken to spread the vaccine and ultimately get rid of smallpox as a threat to public health.

Through the case study, children can investigate the scientific process used to develop vaccines, the value of teamwork and worldwide collaboration in resolving public health issues, and the societal consequences of scientific breakthroughs. By hearing the engaging story of smallpox eradication, children can understand the relevance of science in their everyday lives and how it can be used to improve the world.

BACKPACK
SCIENCES

Child-Friendly Resources

National Geographic Kids

National Geographic Kids provides content on a variety of scientific subjects, such as animals, space, the human body, and the environment. It also offers videos, games, and other activities. A section on recent stories and events is also present on the website.
https://kids.nationalgeographic.com/

Science News for Students

This website offers articles on a range of scientific subjects, such as biology, chemistry, earth science, and physics. Each article is written at an elementary school level and includes a list of resources and activities.
https://www.sciencenewsforstudents.org/

NASA Kids' Club

This website offers games, movies, and other tasks for children that are centered around space and space exploration. Resources like lesson plans and printable exercises are also available on the website for parents and teachers.
https://www.nasa.gov/kidsclub/index.html

KidsHealth

KidsHealth offers essays, videos, and games on a variety of health and wellness-related subjects, such as nutrition, the human body, and mental health. Resources for parents and teachers, including lesson plans and printable exercises, are also available on the website.
https://kidshealth.org/

Wonderopolis

Wonderopolis provides articles and videos on a variety of topics, including science, history, and culture. Each article includes a "Wonder Words" section that explains key vocabulary, as well as related resources and activities.
https://wonderopolis.org/

The Smithsonian Learning Lab

The Smithsonian Learning Lab provides access to millions of digital resources from the Smithsonian's museums, archives, and research centers. Educators can search for resources by subject, grade level, and resource type, and they can also create their own collections and lessons using the resources.
https://ssec.si.edu/stemvisions-blog/where-do-i-find-real-world-science-issues-my-students-can-tackle

Create Science-Based Narratives

Narratives based on science can help children comprehend scientific concepts and ideas. This can involve composing fiction that consolidates logical ideas and thoughts or expounding on genuine logical occasions or revelations.

To introduce the idea of storytelling, use science-based books, movies, and other media as examples. Then, use this science writing lesson to have children tell a story from the main character's perspective. This will help them better understand and apply scientific concepts.

Science Fiction Resources:

The "Picture-Perfect Science Lessons" of the National Science Teaching Association teach science through picture books.

The subject matter of the "Science Comics" graphic novels is science.

BACKPACK
SCIENCES

5 Example Storylines

01 A group of students finds a new planet and must investigate its features, ecosystem, and prospects for habitation using scientific techniques.

02 A new invention from a young inventor has unintended effects, which prompts a scientific inquiry into the device's characteristics and how they impact the area and the people in it.

03 A group of friends establishes a community garden where they must use scientific methods, such as soil science, plant biology, and pest control, to cultivate and harvest their produce.

04 A student comes across a strange new animal species and is forced to conduct scientific research to find out more about its habitat, adaptations, and habits.

05 A team of explorers discovers a secret subterranean world and uses science, including geology, chemistry, and physics, to overcome its obstacles.

Examples of Popular Fictional Narratives to Engage Readers

01 Andrea Beaty's *Rosie Revere, Engineer:* This picture book follows the exploits of a young girl called Rosie who aspires to be an engineer. She learns from her great-great-aunt Rose that perseverance and hard work are necessary for success and that making errors is a normal part of the creative process.

02 Joanna Cole's *The Magic School Bus: Inside the Human Body:* Ms. Frizzle and her class take a wild journey through the human anatomy in this classic book from the "Magic School Bus" series, learning about the various systems and organs along the way. In addition to being educational and entertaining, the tale has plenty of humor and enjoyable illustrations to keep young readers interested.

03 *The Boy Who Harnessed the Wind,* by Bryan Mealer and William Kamkwamba: This motivational true story follows the journey of a young boy named William, who overcomes famine and drought to construct a windmill to provide power for his community in Malawi. The book explores themes of innovation, tenacity, and social justice as well as scientific ideas from engineering, physics, and renewable energy.

04   Connect with Experts

Educators can help children understand how science concepts are applied in the real world and gain a better grasp of how science can be used to solve significant issues by introducing them to experts and professionals. This may motivate children to seek careers in science and show greater interest in their studies. Additionally, introducing children to specialists can aid in the development of collaborative, critical thinking, and problem-solving abilities, all of which are beneficial both inside and outside the classroom.

Inviting Guests Lecturers

You can introduce experts and professionals to the students by inviting them to talk to the class. Visitors can share their insights into their fields, discuss their experiences working on real-world issues, and respond to children's inquiries. To find prospective speakers, educators can get in touch with nearby businesses, universities, and organizations. International organizations like the World Wildlife Fund and Greenpeace also have programs that give students the chance to speak with professionals in different environmental fields.

Virtual Field Trips

Taking children on virtual field trips is another way to introduce them to experts and pros. Today, many businesses many businesses provide webinars or virtual visits that let children learn about what they do. Children could experience a virtual tour of a research center, a farm, or a wildlife sanctuary, for instance. The Jane Goodall Institute, National Geographic, and the United Nations Environment Programme are a few international groups that provide webinars or virtual field trips.

Citizen Scientific Initiatives

These programs enable people, including students, to add to legitimate scientific research by gathering and analyzing data. Children can collaborate with experts and professionals in a variety of scientific disciplines by taking part in citizen science projects, and they can also gain valuable experience conducting real-world research. Additionally, they have the chance to advance scientific knowledge and make significant contributions to discoveries.

International organizations that offer programs for students to connect with experts include:

The Zooniverse

A platform called Zooniverse hosts numerous citizen science initiatives in a variety of disciplines, such as biology, ecology, and astronomy. By identifying and classifying data, children can select a project that suits their interests and contribute to legitimate scientific study.

SciStarter

SciStarter is a platform that links individuals to citizen science initiatives and offers tools to manage and track their contributions. Children can take part in variety of tasks, including tracking the distribution of different bird species and keeping track of the local water quality.

National Geographic

(https://www.nationalgeographic.org/education/student-experiences/)

World Wildlife Fund

(https://www.worldwildlife.org/pages/education)

Greenpeace

(https://www.greenpeace.org/international/act/schools/)

United Nations Environment Programme

(https://www.unep.org/youngenvironmentalawards)

Jane Goodall Institute

(https://www.janegoodall.org/our-work/programs/global-youth-summit/)

Give children opportunities to take action and make a positive impact on the world through their science learning and problem-solving

Encourage children to consider a range of factors, such as the environmental, economic, and social effects of their suggestions, in order to help them develop critical thinking skills about the unintended repercussions of their actions. Children can gain a more comprehensive grasp of the problems and possible solutions.

ASSIGN CHILDREN THE TASK OF CREATING AND CARRYING OUT THEIR OWN INQUIRY TO DEAL WITH A REAL-WORLD ISSUE.

The educator might start by asking the learners to come up with a summary of issues in the real world that they are passionate about. Climate change, pollution, the loss of biodiversity, and access to clean water are a few examples of potential subjects. After the child selects a subject, the educator can collaborate with the child to create an investigation that focuses on a particular facet of the issue.

If the student chooses to focus on pollution, the educator might ask the child to research the effects of plastic waste on nearby waterways. The inquiry may entail gathering water samples from various places, testing them for various pollutants, and then examining the data to spot patterns and trends.

LET CHILDREN SHARE THEIR FINDINGS.

The educator can guide children in planning a presentation after the investigation. The presentation's format depends on the children's and audience's tastes. The children could make a poster or slideshow for a school science fair or a formal talk for a local environmental group or city council.

The educator can help the children structure their presentation, create effective visual aids, and deliver their message clearly and effectively to boost their confidence.

TEACH CHILDREN TO CONSIDER UNINTENDED EFFECTS.

As a final step, the educator can ask the children to consider the effects of their research and presentation. They could examine how their findings could influence policy or how their recommendations could be implemented in their community.

BACKPACK
SCIENCES

COMMON CHALLENGES EDUCATORS FACE WHEN USING REAL-WORLD ISSUES

I JUST DON'T HAVE ENOUGH TIME FOR SUCH A LARGE PROJECT.

Time constraints: Finding enough time to cover all the topics when incorporating real-world issues into science lessons can be difficult. It can be hard to justify spending a lot of time on one real-world issue when there are so many themes to cover.

Solution:

Instead of handling real-world issues as separate units, integrate them into lesson plans. Consider each lesson plan an interdisciplinary lesson. Instead of a whole unit on climate change, an educator can plan in lessons on weather patterns, ecosystems, or human impact on the environment. This can help educators cover all the material while making the content more relevant to children. The educator can add reading comprehension, geography, history, art, writing, speech, math, and more.

I DON'T HAVE ALL OF THE RESOURCES.

Another difficulty educators may encounter is a lack of resources to support their teachings on real-world issues. Real-world problems are frequently intricate and multifaceted, and it can be difficult to locate appropriate learning resources and materials.

Solution:

Utilizing existing resources, such as educational websites, scientific publications, and online databases, is one solution to this problem. Numerous no- or low-cost resources can offer essential information and data to support student learning. In addition, educators can utilize the expertise of professionals in the field by inviting guest speakers, organizing field excursions, or establishing video conferencing connections with experts.

WHAT IF THE CHILDREN AREN'T INTERESTED? THESE TYPES OF PROJECTS USUALLY TAKE A LONG TIME.

Student engagement: Finally, using real-world issues in science classes may make it hard to engage kids. Children may battle to relate to the content or feel overwhelmed by its complexity.

Solution:

Hands-on, inquiry-based learning lets kids examine real-world issues. Educators can help students feel ownership and engagement by letting them create and conduct their own investigations, work in teams, and connect with experts. By connecting the real-world issue to the students' experiences and interests, educators can help kids see the relevance of the topic.

5 QUICK FIXES

Five small, doable suggestions that can lead to quick results for educators or homeschooling parents looking to incorporate real-world issues into their science lessons

1. START SMALL

It's okay to begin by incorporating just one real-world issue into a single lesson or unit. For example, you might choose to focus on the impact of plastic pollution on marine life during a lesson on ecosystems. By starting small, you can gain confidence and gradually build up to incorporating more complex real-world issues into your curriculum.

2. USE VISUAL AIDS

Real-world issues can often be complex and abstract, making it difficult for children to visualize and understand them. Using visual aids, such as videos, infographics, or images, can help to make the issue more concrete and engaging for students.

3. CONNECT WITH EXPERTS

Inviting experts in the field to speak to your class, or connecting with them via video conferencing, can be a powerful way to bring real-world issues to life for your students. Experts can provide firsthand accounts of the issue, answer the children's questions, and offer insights and advice on how to take action.

4. ENCOURAGE TEAMWORK

Working in teams can be a powerful way for children to explore real-world issues and develop critical thinking and collaboration skills. By encouraging teamwork, you can help to foster a sense of ownership and engagement in your students and provide opportunities for children to learn from one another.

5. CONNECT THE ISSUE TO THE STUDENTS' LIVES

Assisting students in recognizing the relevance of the issue to their own lives can be a potent motivator. Consider asking children to generate a list of ways in which the issue affects them or their community, and encourage them to consider potential solutions. By personalizing the issue, you can encourage your students to get involved and make a difference.

BACKPACK SCIENCES

SUMMARY

Real-world issues and problems in science education engage children, provide relevance and meaning, and develop critical thinking, problem-solving, and collaboration skills that are necessary for academic and professional success. Exposing children to real-world problems can create a more inclusive and equitable learning environment that reflects their life experiences, interests, and needs.

Research on using the Montessori method demonstrates how real-world tasks and problem-solving enhance learning. Montessori education stresses real-world applications through hands-on learning, self-directed exploration, and meaningful work. This supports the idea that science education should go beyond memorizing facts and formulas to teach how science affects our daily lives and how it can fix problems.

Real-world issues can be included in science lessons without being an obstacle for educators. For educators seeking to improve the relevance and interest of their science lessons for their students, the five suggestions offered earlier can be a good place to start. Educators can make science more approachable and exciting for children by identifying real-world issues that are pertinent to the child's life and interests, using real-world data and case studies, encouraging teamwork, connecting students with experts, and giving children opportunities to take action.

However, just like with any novel teaching strategy, educators may encounter difficulties when incorporating real-world issues in science classes. For example, a lack of resources, not enough time, and indifference on the part of the students are some frequent problems. Educators can give their pupils a better learning environment by confronting these issues head-on. These issues can be resolved by giving students access to tools, breaking up lessons into manageable chunks, and actively involving them in the learning process.

In conclusion, real-world problems in science teaching engage kids. This develops their critical thinking and problem-solving skills and makes science relevant to all children. Educators will need to acknowledge and create a learning environment that reflects a student's experiences and interests. This helps them succeed academically and professionally.

THE BENEFITS OF INTEGRATING TECHNOLOGY FOR ENGAGEMENT

Why is this important?

There are many Montessori leaders who wonder what Dr. Maria Montessori would think of technology used in the learning environment. We'll never know her true feelings or thoughts on technology as she passed away in 1952 and was unable to experience the technological advancements of today. However, based on her educational philosophy and teachings, one can make some inferences on how she may have felt about technology in the classroom.

Montessori believed in providing a child-centered education, where the focus is on the individual needs and interests of each child. She emphasized hands-on learning and the development of practical life skills. In her view, technology should support, rather than replace, this approach.

Some scholars argue that Dr. Montessori may have embraced technology if it was used in a way that supported her educational philosophy, such as the use of educational software that fostered hands-on learning and problem-solving. Others believe that she may have been more skeptical of technology in the classroom, as she believed in limiting distractions and allowing children to focus on their learning and development through sensory experiences.

Overall, it is difficult to say definitively how Montessori would feel about technology in the learning environment. However, it often comes down to the beliefs of the educator and how they uses it in the learning environment.

This chapter is my attempt to follow Dr. Montessori's philosophy in using technology to enhance science exploration with a hands-on approach. I've included 5 benefits of using technology in the learning environment to engage children and foster a love for science. With each benefit, I've included 3 examples of how to incorporate the benefits into your learning environment.

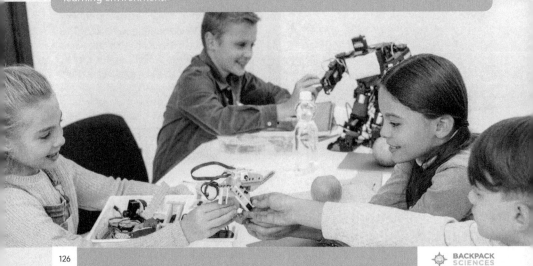

01 Engaging hands-on learning experiences

Children are attracted to technology, a fact any instructor or parent who homeschools their children is likely already aware of. With the help of virtual labs and simulations, for example, technology has made even the most difficult scientific ideas understandable and exciting for students.

To help children visualize difficult scientific ideas, consider using visual aids—for example, videos, animations, and online simulations. These tools can help the subject matter stick in students' minds and come to life. Though I prefer actual field trips, a good virtual field trip is an acceptable option. These online adventures give children a brand-new method to explore scientific ideas.

When it comes to making science more approachable and interesting for kids, using technology in the classroom can be a game changer. Here are a few illustrations:

#1 Virtual Lab Simulation

A Choose a virtual lab simulation that aligns with the lesson you are teaching—for example, a virtual photosynthesis lab.

B Prepare the lesson by setting up the technology you need and making sure that students understand the learning goals for the activity. The learning goal for this activity is to understand the process of photosynthesis.

C Introduce the virtual lab simulation to children and provide guidance on how to use it. Explain that children will be able to conduct experiments and observe the process of photosynthesis in a safe and controlled environment.

D Encourage students to explore the virtual lab and complete related activities and questions. For example, they can change different variables in the virtual lab, such as experimenting with light intensity, and observe how it affects the process of photosynthesis.

E Follow up with a discussion and reflection on what children learned and how they can apply it to real-world situations. Discussion questions could include: What did you learn about the process of photosynthesis? How did changing the light intensity affect the process? Can you think of real-world situations where photosynthesis is important?

#2 Incorporating Educational Videos and Animations

A Choose educational videos and animations that align with the lesson you are teaching—for example, "What is the Solar System?"

B Prepare the lesson by setting up the technology you need and making sure that students understand the learning goals for the activity. The learning goal for this activity is to understand how the Solar System was created and how it is structured.

C Introduce the educational video or animation to the children and provide guidance on what to do next. Explain that the children will be able to visualize the Solar System in an interactive way since we can't really go into space to see how everything is arranged... yet.

D Encourage the children to watch the video or animation and provide opportunities for hands-on learning by having them choose from related activities. For example, they can create a recreation of the Solar System, labelling the planets, moons, and asteroids.

E Follow up with a discussion and reflection on what students discovered and include new projects that scientists are working on right now. Discussion questions could include: What did you learn new about the Solar System? Can you explain the differences between the planets? Can you think of real-world situations where our understanding of the Solar System is important?

BACKPACK SCIENCES

#3 Utilizing Augmented Reality and Virtual Reality Technology

A Choose augmented reality or virtual reality technology that aligns with the lesson you are teaching–for example, the circulatory system.

B Prepare the lesson by setting up the technology and making sure that students understand the learning goals. The learning goal for this activity is to understand the structure and function of the circulatory system in the human body.

C Encourage the children to explore the augmented reality or virtual reality environment. For example, they can identify the different parts of the human body that are directly connected with the circulatory system and their functions, or they can explore more on how the circulatory system interacts with other systems of the body.

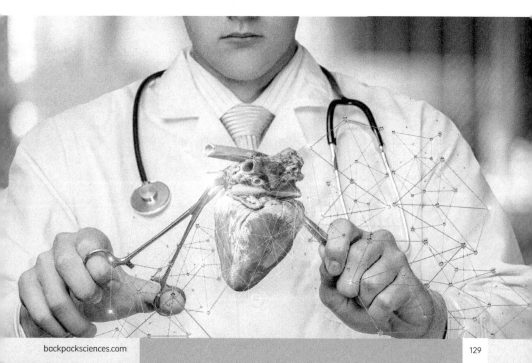

02 Increased accessibility and personalized learning

In terms of science education, technology has the capability to equalize the playing field. Using adaptive learning software, online resources, and assistive technology, all children can achieve success. Adaptive learning software can assist learners in gaining conceptual understanding. From movies to simulations, technology can assist children who require alternative methods to comprehend science. Customization of technology to the needs and interests of each individual makes education more engaging. Technology can ensure that all students receive a science education.

#1 Using Adaptive Learning Software

A Choose an adaptive learning software that is appropriate for the children and your curriculum—for example, plant anatomy and physiology. Don't get too bogged down with age or grade level. Choose one that aligns with your curriculum or topic.

B Familiarize yourself with the software and its features.

C Start to assess your child's starting knowledge and understanding of the topic by asking questions, like whether they know a plant's different parts and their functions.

D Provide the child with personalized learning experiences and allow them to choose activities based on individual interests, strengths, weaknesses, and needs.

E Use the software's reporting and data analysis tools to track student progress and adjust instruction as needed.

F Incorporate discussion questions and activities to help the the child reflect on learning and make connections to real-world applications. For example, based on plants' needs to live and grow, what do gardeners need to make sure they provide and arrange for a healthy, prolific garden?

BACKPACK
SCIENCES

#2 Utilizing Online Resources and Assistive Technology

A Do some research and find the best online resources and assistive technology for your children.

B Familiarize yourself and your childrenwith how to use the technology effectively.

C Give the children opportunities to use the technology and encourage them to explore it on their own time. For example, children could use text-to-speech technology to listen to articles or videos, or they could use assistive tech. to take notes or complete assignments.

D Use the technology to support and enhance your lessons and provide alternative ways to explore different scientific concepts.

E Make sure to incorporate discussion questions and activities to help children reflect on how the technology has impacted their learning. Example discussion questions: How has the assistive technology helped with their understanding of the human body and its systems? How has it impacted their learning in other subjects?

#3 Encouraging Self-Directed Learning

A Provide children access to online resources and technology tools that support self-directed learning. For example, a child might use online resources to study astronomy and space.

B Encourage children to take ownership of their learning by allowing them to work at their own pace and choose topics that interest them. Set some goals—for example, a date for everyone to share their projects.

C Offer guidance and support to help children set and achieve learning goals. For example, how are they going to breakdown the project?

D In addition to in-person check-ins, use technology to provide feedback and assessment opportunities.

E Incorporate discussion questions and activities to help children reflect on their experiences and share their knowledge and insights with others. Discussion question examples: As you became an expert astronomy and space, how do you feel that you were able to make choices and be self-directed?

BACKPACK
SCIENCES

03 Enhanced collaboration and critical thinking skills

Collaboration is key when it comes to science, and technology can definitely play a role. With the help of online tools and resources, children can work together to dive into scientific concepts and ideas, using data analysis and scientific inquiry to develop their critical thinking skills.

With technology, children can collaborate not just with their peers, but also with teachers, organizations, and even individuals around the world. It's amazing how technology can break down geographical barriers and bring people together to achieve a common goal. It's not just about learning science. It's also about developing social and communication skills that are essential for success in not just science but all aspects of life.

As the educator, you could use online tools like "Meet a Scientist" to bring children together and encourage them to research and collaborate with others around the world. Imagine having a virtual field trip to Antarctica to learn about penguins with scientists who are there doing experiments, taking samples and conducting research. The possibilities are endless!

For Example

By utilizing technology this way, children can have a more engaging interactive learning experience. This helps foster their love for science and promotes teamwork and critical thinking skills.

#1 Facilitating Collaboration and Teamwork with Online Tools

A Have children choose a group project or activity that requires them to work together to explore further a scientific concept, For example, they might invest the impacts of different environmental factors on plant growth. The children can work together work to design and conduct experiments to test the impact of light, temperature, and water on the growth of different plants.

B Provide the children with access to online tools, such a Google Docs, to collaborate and share within their groups or with other groups.

C Have children assign themselves roles and responsibilities and share their ideas through the online platform.

D Monitor progress and provide support and guidance as needed.

E Encourage each child to answer questions online or within their presentations: How do different environmental factors impact plant growth?

BACKPACK
SCIENCES

#2 Using Data Collection and Analysis to Develop Critical Thinking Skills

A Select a scientific concept that learners can investigate by gathering and analyzing data using technology. For instance, the children can use temperature sensors and data-logging software to gather information on the temperature and air pressure of various environments, such as indoor and outdoor settings, and then analyze the relationship between the two variables.

B Provide the children the opportunity to work with sensor devices and data-logging tools so they can gather information and make inferences from their observations.

C Plan a session where children can share, talk about, and ask questions about what they learned to hone their critical thinking abilities.

D Possible discussion starters include: How does climate affect air pressure? In what ways did you gather and examine the data? What inferences did you make based on what you saw?

#3

Facilitating Collaboration and Teamwork with Online Tools

A Have the children choose a concept or phenomenon that they can explore through questions (inquiry). For example, they might examine how various types of soil affect plant growth.

B Encourage children to develop experiments to test their theories and ask questions. Children could, for instance, organize experiments to examine how various types of soil affect plant growth and use technology to gather and examine plant height, leaf count, and general health.

C Give children access to technology, such as data-logging software, so they can gather and analyze data from their experiments.

D Plan occasions for children to talk about their research and ask others questions to help them hone their critical thinking abilities.

E Suggested discussion topics: How did you plan your experiment? What aspects did you take into account when selecting your soil types? How did you gather and analyze the statistics? What inferences did you make based on your observations?

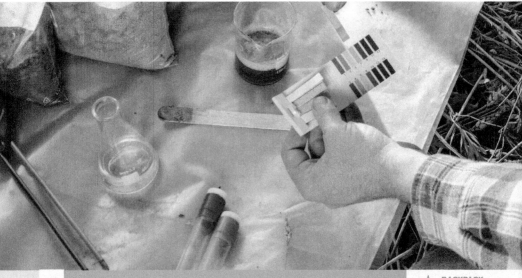

BACKPACK
SCIENCES

04 Access to real-world data and cutting-edge advancements

Technology allows children access to real-world scientific data and the newest scientific advances. This expands science learning, exploration, and discovery. It enhances science comprehension. By keeping children up-to-date with the latest scientific discoveries and developments, we can help inspire and foster a love of science. Real-world data enables children to see how science applies to their lives, making it more engaging and important.

This can also help learners connect science and society in the learning environment and the real world. Exposing children to cutting-edge science inspires them to consider, ask questions, and think critically. We can inspire and equip the next generation of scientists to explore and comprehend the world.

#1 Real-World Scientific Data

A Inform children about real-world scientific data and its relevance in science—for example, data in climate and weather.

B Select an organization—for example, NASA Climate Kids' website—and describe its data and resources. https://climatekids.nasa.gov/

C Let children examine the website's interactive games, videos, and articles.

D Have the children pick a theme or project that uses website data and tools to answer scientific questions or solve problems.

E Children can graph scientific data, such as the Earth's typical temperature.

F Discuss real-world data in scientific study and environmental and climate decision-making with the group.

G Compare NASA Climate Kids' data and information with other sources to assess its truth and reliability.

#2

Staying Up-to-Date
with the Latest Scientific Advancements

A Introduce the concept of staying informed about the latest scientific advancements and explain its importance in the study of science. For example, you may discuss the exploration of different planets and space.

B Direct children to science news, websites, and podcasts. If possible, provide an overview of the resources available.

C Have the children explore the resources and identify a recent scientific advancement that interests them.

D Lead a discussion about the process of scientific research and discovery, including the steps involved and the role of collaboration and teamwork.

E Students will be become experts as they research the specific scientific advancement.

F Allow them to create a way to present their discoveries on the advancement and its potential impact.

G Schedule a time where children present their findings to the class, including a brief summary of the advancements and their potential impact.

Science

Breaking News, World News and Ana

#3 Real-World Scientific Data

A Introduce the concept of connecting with scientists and experts in the field and explain its importance in the study of science.

B Before you begin the lesson, conduct your own research without the children. Look for organizations or universities that connect students with scientists and experts in the field.

C Have a few organizations in mind, but allow the children to explore the available resources and identify a scientist, project or expert they would like to connect with.

D Particular organizations may already have a way to connect, or you can have children write letters. This may include questions about their work, their career path, and any advice they may have for students interested in their line of work.

E Other activities may include a virtual Q & A with a scientist or expert.

F If the children interacted with the scientist—either independently or in small groups—have them create a project that showcases their interactions, such as a video or poster presentation, and present it to the class.

G End this activity with discussions on various careers and opportunities available in the field of science and encourage the children to reflect on their own interests and future goals.

05 Fun and engaging learning

Most children are naturally drawn to technology. If we can convince the children that they are having fun while learning, our job is done! By incorporating gamified learning experiences, virtual field trips, and opportunities for sharing projects and experiments online, we can tap into their affinity for technology and make science both fun and memorable.

#1 Incorporating Games and Gamified Learning Experiences

A First, do a little research on different educational games or activities available online or through a computer program. This may be a specific game that aligns with the topic being covered, or a game that is adaptable–for instance, Jeopardy. Examples of topics include human body, plant growth, and the solar system.

B Use the interactive technology or device to project the game on a larger screen.

C Provide the children with time to play in groups or individually, depending on the structure of the game.

D After playing, facilitate a discussion about what the students learned about the topic. For example, ask, "What did you learn about X that you didn't know before?"

 BACKPACK SCIENCES

#2   Simulated Field Trips

A — Go on a journey or virtual excursion that relates to the subject matter. A digital visit to a museum, zoo, or observatory is an excellent example.

B — Project the virtual field trip onto a larger screen.

C — If feasible, give children headphones or earbuds to improve the sound quality.

D — Stop at different places along the virtual tour to talk about what the children are seeing, answer their questions, and provide information that the learners might not otherwise get.

E — Provide an opportunity for the children to further explore a subtopic they're interested in. This step is essential, as it allows children to dig deeper.

#3 Integrating Technological Elements into Reports and Speeches

A Hopefully, you're on board with the current tendency of letting children pick their own presentations or projects. Art, music, performance, and even poster boards could all fit into this category.

B Provide children with access to electronics so they can share their findings. Possible examples include creating a model of the Solar System, making a digital presentation about an animal they've studied, and devising an experiment to test a scientific theory. Depending on the nature of the assignment, children can work on it independently or in small groups.

COMMON CHALLENGES WITH INTEGRATING TECHNOLOGY

Many elementary teachers struggle with digital integration. Although technology has the ability to improve student's learning experiences, there are several barriers that can make it challenging for educators to effectively utilize it. Lack of access to technology, limited technical expertise, and difficulty integrating technology with the curriculum are the three biggest obstacles to technology integration for educators. Despite these challenges, there are effective solutions that can help educators surmount these obstacles and effectively integrate technology into their lessons to engage and inspire children.

I DON'T HAVE ACCESS TO TECHNOLOGY.

Limited device and internet access are some of the biggest obstacles to incorporating technology into science classes. This makes it hard to give learners multimedia, interactive science lessons.

Solution:

Reach out to local tech firms to ask if they can donate laptops and tablets to schools. Mobile devices for classroom internet access are another option. Groups of homeschooling families can pool resources. Museums, libraries, and workspaces can be used for project time.

I'M TECH-ILLITERATE.

Many educators lack the technical skills to implement technology. This creates difficulty when they want to use devices and software to create engaging and interactive learner experiences. This might make teachers and parents feel inferior.

Solution:

Professional development can show educators how to use technology in their learning environment. Workshops, online classes, and in-person training can help educators integrate technology.

HOW DO I SEAMLESSLY INTEGRATE THIS WITH MY CURRICULUM?

Teachers may face challenges when incorporating technology into their science lessons in a way that supports learning and aligns with the science curriculum. They may struggle to find technology resources that align with their teaching goals and fit into the existing lesson plans.

Solution:

Educators can work with a technology coordinator or curriculum specialist to identify technology resources that support the science curriculum. They can also collaborate with colleagues to share lesson plans and ideas for incorporating technology into their science lessons in meaningful and effective ways.

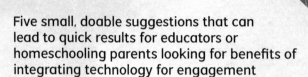

5 QUICK FIXES

Five small, doable suggestions that can lead to quick results for educators or homeschooling parents looking for benefits of integrating technology for engagement

1. BEGIN WITH LOW-TECH OPTIONS

Introducing technology into the learning environment can be intimidating, particularly if you're unfamiliar with the latest gadgets and apps. To begin, attempt low-tech alternatives such as interactive whiteboards, educational games, or videos. These options are straightforward and can be simply incorporated into your lessons.

2. COLLABORATION

Work with another teacher to avoid having to prepare a science lesson every week. This provides you enough time to conduct research, learn, and plan a lesson.

3. USE DIGITAL TOOLS TO SUPPLEMENT HANDS-ON ACTIVITIES

For example, children can use a virtual microscope to observe and analyze specimens, or they can play with forces, motion, and structure using digital simulations.

4. GAMIFY LEARNING

Gamification is an excellent method for engaging learners and making learning enjoyable. Games can be used to reinforce ideas and promote critical thinking. You can, for example, use the popular game Kahoot! to make science quizzes.

5. ENCOURAGE CHILD-LED LEARNING

Give children access to educational websites and apps related to the science subjects they are studying. Encourage learners to make their own digital presentations or videos.

BACKPACK
SCIENCES

SUMMARY

Dr. Maria Montessori's method was to give children a hands-on education that offered them life skills. Many believe Dr. Montessori would have embraced technology if it supported her educational philosophy. Thus, using technology to improve hands-on science exploration can engage children and inspire a love of science.

Technology in the classroom lets educators give learners hands-on, fascinating lessons. Technology can help every learner learn by making the information more accessible and personalized. Online tools and resources can help children work together and build science and life skills.

Technology can also inspire a joy of science by providing real-world data and cutting-edge innovations. Connecting children with real-world data makes science more engaging and important. Finally, gamified learning and virtual field trips can make learning fun and inspire critical thinking in kids.

Technology should assist hands-on learning and practical life skills, not replace them. Educators who follow this theory can use technology to improve learning and foster a love of science. With the right tools and resources, we can inspire the next generation of scientists to explore and comprehend the world.

Chapter 08

OBSERVATION: INCORPORATING A KEY ELEMENT IN MONTESSORI PHILOSOPHY TO ENGAGE CHILDREN

We cannot create observers
by saying "observe," but by
giving them the power and
the means for this observation
and these means are procured
through education
of the senses.
-Dr. Montessori

Montessori education is a pedagogical approach developed by Dr. Maria Montessori that emphasizes individualized learning, hands-on exploration, and self-directed discovery. At the core of the Montessori philosophy is the practice of observation, which involves observing the child's interests, personality, and developmental stage to inspire and engage them in the learning process. In this chapter, we'll discuss the significance of using the Montessori philosophy of observation when teaching children, particularly science.

Observation is a critical skill for Montessori educators to engage children in education. Dr. Montessori stated, "Education is a natural process carried out by the child and is not acquired by listening to words but by experiences in the environment." Experiential learning holds significance in education, and observation helps facilitate such learning experience.

Observation in the Montessori philosophy refers to the process of closely watching and studying the child to gain insight into their interests, needs, and abilities. This enables the teacher to personalize their method of instruction to each child, providing them with the appropriate materials and guidance to support their development. Montessori education uses the fundamental strategy of observation to respond to the child's unique characteristics, preferences, and challenges.

In a traditional, teacher-led classroom, the teacher assumes the role of the primary source of knowledge and direction. The curriculum and teaching methods are typically standardized, with little consideration given to the individual needs and interests of each student.

The Montessori method, meanwhile, recognizes that children learn best when they are actively engaged and interested in the content. The teacher serves as a facilitator, guiding the child through the learning process and providing them with the necessary support and resources to pursue their interests and goals.

The use of observation has particular relevance for science educators. Complex concepts and abstract ideas make learning about science difficult for young learners. However, using the approach developed by Montessori may strengthen science education. By observing the child's interests and preferences, educators can tailor their lessons and the follow-up activities to match the child's individual learning style and provide them with opportunities for hands-on exploration and discovery. This approach can motivate and spark interest for the child in science, improving retention and engagement.

This comparison is supported by researchers comparing traditional and Montessori approaches to education. A meta-analysis of Montessori education research by Lillard and Else-Quest (2006) indicated that Montessori students outperformed traditional students in academic achievement, motivation, and social skills. Rathunde (2016) reviewed Montessori education research and found that it improves academic performance, particularly in arithmetic and language. Schmidt, Harms, and Dueber (2017) found that Montessori students had higher science achievement and attitudes than non-Montessori students. Sander and Sanders (2018) found that Montessori education improved math achievement over time. These findings suggest that implementing the Montessori method of observation into scientific instruction can improve student learning and science attitudes.

The use of observation is a valuable tool for science educators seeking to create a more engaging and effective learning experience for elementary children. By observing the child's interests and personality, educators can tailor their approach to meet the child's individual needs, providing them with opportunities for hands-on exploration and discovery. Using observation as a foundation can lead to deeper engagement, conceptual comprehension, and a more positive attitude toward science.

Personal Story

I had been teaching for about a decade in an Upper Elementary Montessori classroom, and I had always found that the Montessori philosophy of observation helped me to connect with my students on a deeper level.

One year, I was teaching a lesson on photosynthesis. I noticed that one of my students, Jenny, was having trouble grasping the concept. It really is an abstract concept, and at best, some of the features can be shown microscopically or through a simulation video.

Jenny was a bright student, but she was struggling to understand the process of how plants convert sunlight into energy. I was guiding the students through the follow-up activities for the photosynthesis lesson in the Backpack Sciences Membership, including:

- Testing how the color of light affects growth
- An experiment altering soil, water and sunlight
- Watching a music video on photosynthesis
- Creating a comic strip

But I noticed that Jenny still didn't appear to understand the concept of HOW photosynthesis happens. Putting my observation skills to use, I took a more hands-on approach to teaching the concept of photosynthesis.

I set up an experiment in the classroom using spinach leaves and a light source. I asked Jenny to observe the leaves for a few minutes and then make note of any changes she noticed. After a few minutes, Jenny excitedly pointed out that the leaves had turned slightly darker in color. I explained to her that the leaves had absorbed sunlight and were beginning the process of photosynthesis.

With this simple demonstration, Jenny was able to understand the basic concept of photosynthesis. I was able to engage Jenny by using my observation skills to inspire her and make the scientific concept memorable. From that day on, Jenny became more interested in science and asked more questions during class.

By using the Montessori approach of observation and hands-on exploration, I was able to connect with Jenny and help her understand a challenging scientific concept. This approach not only helped Jenny learn, but also inspired her to develop a deeper love for science. I know that this experience will have a lasting impact on Jenny's education and am grateful for the opportunity to use my Montessori observation skills to help children to succeed.

5 Strategies
to Implement Observation

 Provide open-ended materials

One strategy is to provide open-ended materials for children to explore. Observing children's use of these materials can reveal their creativity and problem-solving abilities.

For instance, an educator may notice that a child is proficient at building with blocks, Legos, or recyclable materials. The teacher can then help the child design and build a bridge to further discover engineering concepts. This technique encourages creativity and independence in children.

#1 Adding Versatility

A Choose a science topic—for example, the states of matter.

B Select open—ended materials —Choose open-ended materials that will allow exploration in a hands-on way. This may include the use of ice, water, steam, and containers of various sizes.

C Set up the materials - Encourage inquiry and experimentation. Arrange containers so children can pour and mix the different states of matter. Encourage children to make predictions, test their hypotheses, and record their findings in their scientific notebooks.

D Observe and document - Oftentimes, the best observations are taken when you don't interrupt. Take notes on their interactions and questions to you or their peers. Use your observations to gather information, guide your teaching and provide support where needed. Take pictures or videos of your students' experiments and observations.

E Encourage reflection and discussion - After exploration, gather children to discuss their experiences. Ask "What surprised you?" and "What did you observe?" Encourage children to discuss their results and hypotheses. Use this time to explain misconceptions or reinforce essential concepts.

 When provided with open-ended materials, learners can experiment with science concepts. This approach encourages curiosity, creativity, and critical thinking. As an educator, you can observe the child's learning to guide your instruction and provide concentrated support.

 BACKPACK SCIENCES

Providing Choices

#2

By providing children with choice in their learning, we can observe which activities and topics each child is naturally drawn to. Educators can tailor their lessons and activities to better engage each child.

For example, an educator may observe that one child is particularly interested in animals while another is interested in plants. The teacher can then provide activities that allow each child to explore their interests further, such as by researching and presenting on their chosen topic. This strategy benefits the educator by promoting a sense of ownership and investment in learning among children.

A Design individualized follow-up activities – For this example, let's use the concept of photosynthesis. The follow-up activities will encourage the children to explore the concept of photosynthesis in different ways. For example, you might design activities that focus on the chemical reactions involved in photosynthesis, the role of chlorophyll, or the environmental factors that affect the process.

B Provide choices –Offer the children a variety of choices for follow-up activities that allow them to explore the concept of photosynthesis in ways that match their interests and learning style. You can offer different activities at different levels of complexity and challenge to support all of your students.

C Support student choice–As children make their choices, provide support and guidance to help them succeed. Offer resources such as books, articles, and videos to help students understand the concept better.

D Observe and document - Observe the children as they work on their chosen activities. Take notes on their interactions and the questions they ask. Use your observations to guide your teaching and provide support where needed. Take photos or videos of your students' experiments and observations, if possible.

Providing choices in follow-up activities is important because it allows children to take ownership of their learning and pursue their interests. By giving children a variety of options, you encourage them to think critically about the concept they just learned and apply it in different ways. This approach helps to engage with the material on a deeper level, which can lead to more meaningful learning outcomes.

From the teacher's perspective, offering choices in follow-up activities helps to develop observation skills by allowing you to see how the children approach the concept of photosynthesis in different ways. By observing their choices and interactions, you can gain insight into their interests and learning styles, which can inform your future teaching strategies. Additionally, providing choices allows you to differentiate your instruction and provide targeted support where needed, leading to a more effective and inclusive learning environment.

Encouraging Exploration

#3

Educators can use observation skills to encourage children to explore their interests and passions. By observing what children are naturally drawn to, teachers can tailor their lessons and activities to incorporate those interests.

For example, after children study a habitat, the educator might ask the child more about the habitat and about the other plants and animals that live in this habitat. This can also go one step further with environmental issues that may cause extinction or force animals to adapt to climate changes or human influences.

A Choose a science concept - The first step is to choose a science concept that you want your students to explore. For this example, let's use the concept of the water cycle.

B Spark curiosity - During the lesson, spark curiosity in the children by encouraging them to ask questions and share their observations. Ask them what they notice about the water cycle and how it works.

C Offer resources - After the lesson, offer the children a variety of resources such as books, articles, and videos to help them explore the water cycle in more detail. Provide them with guidance on where to find reliable and age-appropriate information.

D Encourage exploration - Encourage the children to further explore the water cycle by conducting their own experiments, creating models, or researching related topics. Depending on the children, you may have to do this in phases. For example, if you're doing this for the first time, the children will need more structure in what "exploration" means. You'll have to give them details on your expectations, including time, topics, and activities. As the children get more accustomed to this open-ended opportunity, it'll become less work and preparation for the educator. Allow them to pursue their interests and offer support and guidance as needed.

E Observe and document - Observe the children as they explore the water cycle further. Take notes on their questions, observations, and experiments. Use your observations to guide your teaching and provide support where needed.

By allowing and encouraging further exploration of the water cycle after the lesson, you can utilize your observation skills to engage the children in loving science. By offering resources and encouraging exploration, you can support their interests and guide them to deeper levels of understanding and independency. Additionally, by observing and documenting their progress, you can gain insight into their interests and learning styles, which can inform your future teaching strategies.

BACKPACK
SCIENCES

#4 Conducting Weekly Individual Conferences – Benefits and Solutions

A Conducting one-on-one discussions with children is one of the best techniques to use observation skills. Educators may observe and assess each child's learning preferences and interests during these weekly conferences and adjust their lesson plans accordingly.

B An educator might notice, for instance, that a child struggles with math but does well in painting. In order to make math lessons more interesting for the child, the teacher might then add art into the teaching material. The educator benefits from this approach as it helps them better grasp the unique needs and strengths of each child.

Schedule one-on-one time during the child's individual work time. For example, an educator may provide each child with 30 minutes a week to work on a project while providing individual help and feedback. In a Montessori learning environment, this may be during the three-hour morning work cycle.

When I was teaching in the classroom, I enjoyed scheduling one-on-one check-ins on Fridays. I used this time to see how their week went, and we would also set goals for the following week.

Incorporating one-on-one time into the classroom routine is another option. An educator may hold a weekly "check-in" where each child discusses their success and challenges. Depending on the child's comfort, this can be done alone or in a group.

Although teachers may struggle to find the time and resources for one-on-one conferences, I believe the benefits outweigh the difficulty. Educators may help learners succeed academically and emotionally by creating close relationships, personalizing curriculum, and offering personalized support and feedback. One-on-one conversations with children can also inspire instructors and remind them of their impact on young lives.

Montessori Connection

The Montessori philosophy, which emphasizes the value of personalized attention and individualized learning experiences, encourages the scheduling of weekly one-on-one time with children. Finding the time and resources to provide each child with individualized attention can be an issue for many educators. But there are many advantages to scheduling regular one-on-one time with children, and the work put into this practice can benefit both the learner and the teacher.

One-on-one time with children allows educators to build a strong relationship with each child. When an educator spends time with a child one-on-one, they can learn about the child's interests, learning style, and individual strengths and weaknesses. This knowledge allows the teacher to customize instruction and provide individualized support to help the child thrive.

One-on-one time allows educators to engage in deeper conversations about a child's progress and challenges. A child is more inclined to express themselves to their teacher when they know they have dedicated time with them each week. This can provide valuable insight into the child's learning and emotional needs, which can guide an educator's approach to instruction and support.

Despite the benefits, many educators lack the time and resources to give students one-on-one attention. Lesson planning, grading, and classroom management often make it challenging to schedule individual time. However, there are various ways educators can maximize one-on-one interaction with children.

#5 Using Group Activities

A Through group activities, educators can learn about each child's social skills and learning style through observing how they interact and collaborate.

B An educator might notice, for instance, that a child naturally takes charge of a group project. The teacher may then provide the child an opportunity to take the lead during subsequent activities, which will increase their self-assurance and sense of accountability. However, via observation, the teacher may recognize a child who may benefit from being given the opportunity to take the lead in the activity. This method is beneficial to the teacher because it encourages good relationships between children and a sense of community in the classroom.

BACKPACK
SCIENCES

STRATEGIES

COMPREHENSION DEVELOPMENT

Solution:

In order to thoroughly investigate and comprehend a subject like science, cooperation and teamwork are often essential. Group science activities are an excellent way to get kids collaborating while exchanging their observations in the learning environment, which can help them fully understand science concepts more effectively. Additionally, group scientific activities may help educators in gathering essential information regarding the child's interests and learning preferences, which can guide the next lesson and activities.

GROUP DYNAMICS

Solution:

Managing the dynamics of the group is one challenge educators might face when introducing group science activities. Children may participate in the activity to varying degrees of interest or participation, or there may be conflicts that develop between group members. Establishing clear expectations and ground rules for group work – such as taking turns speaking, paying attention to one another, and valuing one another's thoughts and contributions – could assist with tackling this issue.

PINPOINTING AREAS OF IMPROVEMENT

Solution:

Group scientific exercises also give educators a chance to pinpoint each student's unique skills and areas for improvement. For instance, a group task or project may demonstrate a child's capacity for original thought or problem-solving, or it may show areas where a child may require additional support or direction. Educators can use this information to develop future lessons and activities that are tailored to the interests and strengths of their pupils.

#4   *LACK OF PARTICIPATION*

Solution:

An excellent technique to engage kids who might be reluctant to participate in individual activities or conversations is through group scientific activities. In a group scenario, where they may bounce ideas off of other children and receive feedback and encouragement, some children might feel more at ease sharing their thoughts and observations. Teachers may promote a supportive learning environment and help students develop a feeling of community by giving them opportunities to collaborate.

#5   *USING A CHALLENGE*

Solution:

One example of a group science activity that can help develop observation skills is a science challenge or competition. For example, you may have heard how a teacher may challenge groups of children to design and build a device that can protect an egg from a high fall. The groups are given a set amount of time to plan, design, and build their device, and then they test their device by dropping it from a predetermined height. The group whose egg survives the fall intact is declared the winner.

This activity encourages the children to carefully consider the properties of the materials they are using and the forces that will be acting on their device during the fall. Additionally, it can help students learn how to work collaboratively and share their ideas and observations with their peers.

02 Questions to ask and take notes on when observing

During group activities, many science teachers focus on the science concepts. But I'd like you to think about how group activities help you develop your observation skills. Here are some questions to consider as you observe the students' collaboration:

· Who is speaking during discussions?
· Who is listening?
· How does the "leader" of the group take in all of the suggestions? Or do they?
· What is the communication style of the group?

#1 Examples of Observations

A During a science challenge activity in my Montessori classroom, the children were asked to design a bridge that could hold a certain weight using only popsicle sticks and glue. One student, Maria, came up with a unique design that utilized triangles for extra support. Her design was successful and held the weight with ease, while other groups' bridges collapsed. This challenge highlighted Maria's creativity and problem-solving skills, and allowed her to shine in front of her peers.

B In another group activity, the students were asked to create a simple machine using everyday materials. One student, Jack, struggled with the activity and was unable to come up with a functional design. This group activity allowed me, as the teacher, to observe Jack's need for additional support and guidance in understanding simple machines. I was able to provide Jack with individualized instruction and scaffolded learning opportunities in the following weeks to help him better understand the concepts.

C During a class discussion, a shy student named Alex was hesitant to share his thoughts and opinions on a particular topic. However, during a group project, Alex was assigned a leadership role and provided with feedback and encouragement from his peers. This helped him gain confidence in his abilities and eventually led to him feeling more comfortable sharing his ideas individually in class.

Group activities enabled students to demonstrate their abilities, identify areas where they need support, and provide opportunities for peer comments and encouragement. As an educator, it helped me discover each student's learning style and needs to deliver individualized support and instruction. Group activities foster classroom community and teamwork, which is essential to a positive learning environment.

Group science activities may engage children and improve their observation skills. Clear expectations and ground rules can help educators manage group dynamics and create a productive, supportive learning environment. By providing opportunities for children to work together, educators can also gather important information about their students' strengths and interests, which can inform future lessons and activities. Ultimately, group science activities can help foster a sense of community and encourage a lifelong love of science in children.

Reflection

#2

Many skip over taking the time to reflect after their observation. Reflection enables people to acquire insights from their experiences. Reflection helps Montessori guides develop and process what was observed of children. In the context of science education, reflection allows children to connect their experiences to scientific concepts.

A Educators struggle to find time and space to reflect. Reflection is essential to further develop learning and to comprehend children's thinking and problem-solving. One suggestion is to schedule daily reflection time, such as after the lesson or at the start of the day. Depending on the children and scheduling, this can be done individually or in groups.

B Many science activities may encourage reflection. For example, after a science experiment, children are encouraged to analyze the results. They can also discuss their thinking processes and strategies for problem-solving. Additionally, journaling and drawing activities can be used to encourage deeper reflection and understanding.

C Reflection has several benefits for children and educators. Reflection helps children develop critical thinking, problem-solving, and a deeper understanding of scientific concepts. Additionally, it identifies where they may need further support or guidance.

D For the educator, reflection provides valuable insights into each child's learning and development, which can inform future lesson planning and individualized instruction. It also identifies areas in which educators may need to modify their approach or provide additional guidance.

E Personal reflection is an effective way for educators to develop their observation skills and gather information about a child's learning and development. An educator can review a recent science lesson and determine how each child engaged with the subject and what strategies promoted deeper learning. They can also reflect and improve their teaching practices and identify areas for improvement.

F There are various ways to incorporate reflection into science lectures. For instance, one can use open-ended questions to encourage children to reflect carefully on the subject matter and their personal experiences. Additionally, they can provide children the chance to explore ideas with their peers, which can result in insightful discussions.

To sum up, reflection is an essential tool for Montessori guides to improve their observational abilities and understand more about a child's learning and growth. Making time for reflection is crucial if you want to promote deeper learning and acquire an understanding of how children process information and address problems. It helps teachers promote critical thinking, problem-solving abilities, and a deeper knowledge of scientific principles.

BACKPACK
SCIENCES

03 Observation activities

Silent Observation Game

#1

In the 'silent observation' technique, children concentrate their attention on a specific object or phenomena without any interruptions. Montessori educators use 'silent observation' to help children develop observation skills, mindfulness, and a feeling of surprise and curiosity. This method can be adapted to different ages and interests to help children understand about the natural world and how science is all around them. This method can be used to improve children's understanding of scientific ideas, cultivate mindfulness, and foster a sense of wonder and curiosity.

A Choose an object or phenomenon for children to observe. This could be a plant, an insect, a cloud, or any other natural element that is available in the learning environment.

B Explain to children that they will be engaging in silent observation, which means that they will be focusing their attention on the object or phenomenon without speaking or making any noise.

C Invite children to sit quietly and comfortably in front of the object or phenomenon. Encourage them to relax their bodies and to breathe deeply.

D Ask children to focus their attention on the object or phenomenon and to observe it carefully. Encourage them to notice any details, such as its color, texture, shape, and movement.

E Encourage children to use all their senses to observe the object or phenomenon. For example, if they are observing a plant, they could observe its color and texture, as well as its scent and the sound of the wind rustling its leaves.

F Allow children to observe the object or phenomenon for a few minutes, depending on their age and attention span.

G After the observation period, invite children to share their observations with the group. Encourage them to describe what they noticed and to ask questions about what they observed.

H Use this opportunity to guide children's exploration of the scientific concepts related to the object or phenomenon they observed. For example, if they observed a plant, you could discuss the process of photosynthesis and the role of sunlight and water in plant growth.

I Repeat this activity with different objects or phenomena, and encourage children to develop their observation skills by focusing on different details each time.

Using Scientific Tools and Instruments

#2

Montessori instructors can help children understand scientific concepts by using scientific equipment and instruments. This approach may enhance children's hand-eye coordination and fine motor skills. Scientific instruments allow children to make precise and detailed observations of the nature world. Children can view the world differently and understand scientific concepts differently.

A Let learners explore a scientific tool. This could be a magnifying glass, microscope, balancing scale, or other tools suitable for their age and interests.

B Show children how to use the scientific tool to observe and measure nature.

C Show children how to use the scientific tool or instrument and invite them to explore it themselves.. Encourage them to handle the tool properly and follow safety instructions.

D Give children a range of natural objects to measure using the scientific tool—for example, leaves, insects, and rocks.

E Demonstrate how to take careful, detailed notes in their scientific notebook.

F Guide children's scientific inquiry of the natural objects they observed—for example, after the children have observed an insect using a magnifying glass, discuss the insect's anatomy and behavior with them.

G Repeat this practice with other scientific tools and natural objects to help children improve observation skills and gain a deeper understanding of scientific concepts.

Educators can take observation notes using several Montessori teaching practices and data collection techniques. Educators can recognize each child's progress, abilities, and challenges by employing multiple methods. These insights can inform class planning, progress tracking, and parent communication.

BACKPACK
SCIENCES

Different Observation Collection Strategies and Tools

Observation Checklist

Checklists allow educators to record a child's behavior, skills, and milestones during the lesson or day. Checklists can evaluate the child's progress and identify areas for improvement. https://montessoricompass.com/ provides a comprehensive online record-keeping system for Montessori schools, which includes observation checklists for tracking student progress.

Anecdotal Records

TrilliumMontessori
.ORG

Anecdotal records are brief accounts of a child's actions. These narratives might document the child's progress, interests, strengths, and concerns. Anecdotes can help teachers design lessons, evaluate them, and connect with parents. Trillium Montessori has wonderful examples of different record keeping systems, https://www.trilliummontessori.org/evolution-of-my-montessori-record-keeping/

Video Recording

Video recording can be a useful tool for observing children's behavior during lessons. By recording lessons and reviewing them later, teachers can identify areas for improvement, see how children interact with each other, and analyze their teaching style. This technique can be used in any subject area and is particularly useful for tracking progress in sensorial, math, and science lessons. **If you're doing this in a classroom, teachers need written consent from the parents to allow videotaping of the children.

Running Records

Running records are a type of observation note that captures what the child is doing, saying, and thinking during a specific activity or lesson. This technique is commonly used in language lessons to track progress in reading and writing skills. An example of how to use running records can be found here: http://www.scusd.edu/sites/main/files/file-attachments/decs_running_records_australia.pdf This technique is best used in language and cultural subjects.

Mind Maps

Mind maps can be used to organize and analyze the observations made during a lesson or over a period of time. They can help teachers identify patterns and connections between different behaviors and skills. An example of how to use mind maps for teaching can be found here:
https://www.edrawsoft.com/mindmap-usage-for-teaching.html

SUMMARY

Observation is the act of paying attention to details and gathering information through one's senses. It's vital to scientific investigation and can be learned. During Montessori teacher training, instructors emphasize the use of observation to help inform teachers how children explore and discover the world around them. As educators establish connections and draw inferences from their observations, they develop critical thinking and problem-solving skills in order to provide the best learning environment for the learners.

This is also important for the children in the Montessori learning environment. Children are encouraged to engage in scientific inquiry through observation. They are encouraged to explore nature and record their observations. Observation helps children grasp scientific topics.

Observation also plays a vital role in the development of children's language skills. The Montessori approach to learning language encourages the development of oral language skills before the written language. An example of this is encouraging children to describe their observations and experiences. This helps the children develop effective communication and language skills.

Observation is essential for children's social skill development. Children gain an understanding and respect for other people's viewpoints and emotions by watching others. Additionally, they can learn the value of collaboration and excellent communication, two abilities that are crucial for both personal success and scientific research.

Children benefit from observation in a variety of ways, including the promotion of mindfulness, a sense of wonder and curiosity, and the facilitation of scientific inquiry.

Montessori instructors aid in the development of children's intrinsic sense of wonder and curiosity by encouraging them to notice and investigate the world around them, sometimes by observing silently. This sense of wonder and curiosity can motivate children to keep exploring and learning throughout their lives.

Numerous methods are employed by Montessori educators to promote children's observation skills. One such strategy is "silent observation." With this method, children are urged to focus on a certain object or phenomenon without any interruptions. They are able to concentrate on the details and hone their observational skills as a result.

The use of scientific equipment and tools by educators is another observation method. Educators encourage children's closer observation and exploration of the natural world by offering and encouraging exploration with microscopes, magnifying glasses, thermometers, and other scientific equipment. This fosters children's hand-eye coordination and fine motor abilities as well as facilitating scientific inquiry.

In order to engage children in science instruction, educators have to develop the skill of observation. In addition to supporting the growth of children's scientific inquiry abilities, language abilities, social skills, and sense of wonder and curiosity, observation encourages children to explore the natural environment. Observation helps children establish connections and reach conclusions. Educators may encourage children to observe through a variety of methods, including quiet observation and the use of scientific equipment. When educators place a strong emphasis on the value of observation in education, children are better equipped for a lifetime of learning and exploration. Observation is crucial for both scientific research and everyday life.

BUILDING COMMUNITY USING SCIENCE

Building community in learning environments is a critical component of providing children with a positive and engaging learning atmosphere. Children who feel connected to their peers and teachers are more likely to be motivated, engage effectively in class, and achieve academic success.

Studies show that a positive learning climate increases academic achievement, attendance, and reduce negative behaviors such as bullying (Bradshaw, Waasdorp, & Leaf, 2012). This positive environment can be created by deliberate efforts to foster a feeling of community among students and teachers. Educators can foster child engagement and scholastic success by developing a culture that values respect, empathy, and collaboration.

Science activities and content offer educators an excellent chance to foster community among their pupils. Science has the potential to arouse children's curiosity and wonder, and when taught collaboratively and interactively, it can promote a sense of community and shared learning experiences. Furthermore, science is a topic that allows children to explore and connect with their personal interests, values, and identities.

Incorporating culturally relevant science lessons and follow-up activities can engage children and build community in their learning environments. For example, educators can encourage children to explore scientific concepts related to their cultural heritage or personal interests. They can also encourage children to share their scientific knowledge with classmates to foster a more inclusive and diverse learning atmosphere.

Science activities that emphasize teamwork and communication can also help children develop positive social skills and behaviors. Group projects, lab experiments, and peer review activities encourage children to collaborate and communicate in order to reach a mutually beneficial goal. Such duties promote cooperation, learning, and positive relationships.

Building a sense of community among children is crucial for creating a positive and engaging learning environment. With the right tools, educators can use science as a natural segue to foster community and student engagement. Science activities that are culturally relevant, meaningful, and supportive of teamwork and communication can foster a sense of community in children, increasing motivation, participation, and academic success.

BACKPACK SCIENCES

Montessori

Montessori stresses community as an aid to social and emotional development. To achieve this, the learning environment should be a microcosm of the world and should encourage children to learn about different cultures and perspectives. These learning environments facilitate and teach children of different ages and stages. Children learn conversation, collaboration, and community.

Cultural activities help children comprehend and appreciate different countries, traditions, and perspectives. Educators should include activities involving music, art, and food to teach history, geography, and society. These activities help children honor diversity and respect others.

Studies show that the Montessori method helps children's social and emotional growth and community formation. Lillard and Else-Quest (2006) discovered that children's social and emotional development was better in Montessori classrooms than in conventional classrooms. Positive social behaviors, such as assisting and sharing, were found to be more prevalent among Montessori students.

Lillard's research into Montessori nursery graduates continues. According to Lillard et al. (2017), children who are exposed to the Montessori method are more likely to work together and share ideas. Increases in both intelligence and mental maturity can be attributed to the Montessori method.

Rathunde and Csikszentmihalyi (2005) discovered that children's intrinsic motivation and engagement in learning were significantly greater in Montessori classrooms than in conventional classrooms. The authors postulate that the Montessori method's focus on cultivating a sense of community and fostering the growth of students' social and emotional competencies may be at least partially responsible for this finding.

Cultural responsiveness is crucial in Montessori classrooms, according to Cohen (2019). The author claims that a culturally responsive approach that values students' backgrounds can foster a feeling of community by providing an inclusive learning environment. Cohen believes that cultural studies might help establish community by teaching learners about and appreciating diverse cultures.

According to Montessori education, building community in the classroom helps children develop socially and emotionally. Cultural studies and multi-age classes teach students to value differences and respect others. The Montessori method has been proven to promote community and interest youngsters in learning.

10 Strategies to Build Community

 01 Establish a welcoming and inclusive classroom environment

Science education helps children understand the natural world and develop scientific inquiry skills. However, many children lack resources, encounter science barriers to access high-quality science education, or have limited hands-on learning opportunities.

Overcoming these challenges and engaging all learners in science requires a welcoming and inclusive learning environment. Creating a positive learning atmosphere that encourages differing viewpoints and experiences allows children to ask questions, take chances, and explore scientific concepts. Children are more likely to appreciate science and regard themselves as valuable scientists in such an environment.

The Science Behind the Recipe

 #1

Have children bring in a cultural food item and investigate the science behind the recipe. This helps build community by encouraging children to share their cultural traditions and learn from each other. By investigating the science behind the recipe, children can also learn about the chemical reactions that occur during cooking and the nutritional value of different foods, which can promote scientific literacy and healthy eating habits.

1 Start by explaining to children the objective of the activity, which is to explore the science behind the recipe of a culturally significant food item.

2 Encourage each child or pair of children to select a cultural food item to investigate.

3 Children can research the cultural significance of the food item, as well as the history and ingredients of the recipe.

4 Children can then experiment with the recipe and observe the chemical reactions that occur during the cooking or baking process.

5 After the food item has been prepared, children can present their findings, discussing the scientific principles involved in the recipe and how it relates to the culture it originated from.

Class Mural

Creating a class mural that highlights the diversity of the classroom, including different scientific discoveries from different cultures, can help build community by celebrating the unique backgrounds and perspectives of each child. By including scientific discoveries from different cultures, children can learn about the contributions of people from around the world to scientific knowledge and appreciate the interconnectedness of science and culture.

Begin by introducing the concept of a class mural and explaining that it will be used to celebrate the diversity of the classroom, including scientific discoveries from different cultures.

1

2

Brainstorm ideas for the mural, including different scientific discoveries and cultural symbols..

Once a plan for the mural has been established, students can work together to create the mural using materials such as paint or markers.

3

4

As the mural is being created, educators can facilitate discussions about the different cultural and scientific elements being included.

Once the mural is complete, it can be displayed prominently in the classroom as a reminder of the diversity and inclusivity of the classroom.

5

Scientific Investigation on Skin Color

#3

Conducting a scientific investigation on skin color to demonstrate that differences in skin color are due to variations in the amount of melanin in the skin can help build community by promoting understanding and acceptance of differences. Children can learn that skin color is determined by genetics and that differences in skin color have evolved as a result of environmental factors such as exposure to sunlight. By understanding the science behind skin color, children can appreciate the diversity of human populations and develop empathy and respect for people who look different from themselves.

Begin by introducing the concept of skin color and discussing the factors that influence it, including genetics and exposure to sunlight.

1

2

Children can then research the scientific principles behind skin color, including the role of melanin in the skin.

The children can then conduct a scientific investigation, such as measuring the amount of melanin in different skin samples using a spectrophotometer.

3

4

As the investigation is being conducted, educators can facilitate discussions about the different factors that influence skin color and the impact it has on society.

After the investigation is complete, children can present their findings to the class and discuss the implications of the scientific principles involved.

5

Adaptations

Investigating how different animals have adapted to their environments, and discussing the different adaptations seen in different cultures, can help build community by promoting an understanding and appreciation of diversity. Children can learn that different animals have evolved unique adaptations to survive in different environments, and that people from different cultures have also developed unique ways of adapting to their environments. By recognizing and celebrating the diversity of adaptations in animals and cultures, children can develop a sense of curiosity and respect for the natural world and for people from different backgrounds.

Begin by introducing the concept of adaptation and explaining how different animals have adapted to their environments over time.

1

2

Children can then research different animals and their adaptations, including those found in different cultures.

As a group, discuss the different adaptations and how they relate to the culture and environment in which they were developed.

3

4

To reinforce the concept, educators can have children create visual representations of the different adaptations they have learned about.

To conclude the activity, children can share their visual representations and discuss the different adaptations seen in different cultures.

5

Health Outcomes and Health Equity

Different cultures and people have varying health outcomes and strategies when it comes to health equity. Children can learn that social and economic issues, such as race, ethnicity, and finances, can have an impact on health outcomes—for example, access to healthcare, nutrition, and environmental quality. Children can develop a sense of social responsibility and empathy for those who may be experiencing health inequalities by recognizing these differences and suggesting solutions.

Introduce the idea of health equity and discuss how many cultures and people may have gaps in their health.

Separate the children into smaller groups, and let them each pick a different city or culture to research. Give children access to materials they can use for their research, including books, papers, and websites.

Suggest each group explain its results, and then lead a discussion on the shared and distinctive characteristics of the various cultures and communities.

Encourage children to consider their own experiences with health inequalities and to share any unique thoughts or tales they may have.

Give children follow-up tasks or activities that let them continue learning about health equity and take initiatives to advance it in their own communities. This can entail carrying out community service projects that advance health equity or writing essays or making speeches on the subject.

After that, children can look at the variables that affect various health outcomes, such as socioeconomic status, healthcare access, lifestyle choices, and prejudice.

Give each group the assignment of researching the health results of the culture or community they have chosen, as well as any discrepancies or inequities they may find. Encourage learners to consider these discrepancies' contributing factors critically.

Come up with ideas for promoting health equity within each of the cultures and groups under study, as well as for broader societal advancement.

Summarize the conversation and stress the value of ensuring health equity for all people, irrespective of their background or culture.

02 Encourage students to introduce themselves to one another

Building a sense of community is a crucial component of creating a welcoming and inclusive learning environment. Encourage children to introduce themselves so they may get to know each other, create relationships, and feel like they belong in the learning environment. This can help children feel more comfortable discussing, working in groups, and asking for help from their peers and instructors.

Questioning

#1 By asking a question, the educator places a focus on on the discussion. Choose a question that all children can answer regardless of socioeconomic class, experiences, gender, culture, or religion.

Allow children to introduce themselves during the first week of school.

1

2

Ask children, "What is one thing you are passionate about?" or "What is one fun fact about you?"

Encourage children to actively listen to their peers' introductions and ask follow-up questions.

3

Scientific Selfie

Have children take a class "scientific selfie" that includes a photo and a brief description of their scientific interests and hobbies. This activity can help students learn more about each other's interests and backgrounds related to science, while also promoting a sense of community and connection within the classroom.

Explain the activity to the children and provide examples of what a "scientific selfie" might look like.

Allow time for children to take a photo of themselves and write a brief description of their scientific interests and hobbies.

Invite children to share their "scientific selfies," either by displaying them on a bulletin board or presenting them to the group.

Science Speed Dating

Conduct a "science speed dating" activity in which children rotate and interview each other about their scientific interests and background. This activity allows children to get to know each other's scientific interests and background, while also practicing communication and active listening skills.

Divide children into pairs and provide a set amount of time (e.g. 2-3 minutes) for each pair to interview each other about their scientific interests and background.

After the time is up, have children rotate and find a new partner to interview.

Encourage children to actively listen to their partners and ask follow-up questions to learn more about their interests and background.

BACKPACK
SCIENCES

Board Game

#4

Have children design and create a science-based board game that highlights different scientific concepts and encourages collaboration and communication. This activity allows children to work together in groups to create a fun and interactive way to learn about different scientific concepts, while also practicing collaboration and communication skills.

Divide children into groups and encourage them to create their own set of guidelines and materials to create a science-based board game.

1

2

Encourage children to work together to decide on the theme and scientific concepts to include in their game.

Invite each group to present their game to the class and play each other's games.

3

Contributions of Scientists

#5

Have children work in groups to research different scientists who have made significant contributions to science and present their findings to the class. This activity allows children to learn about different scientists and their contributions to science, while also practicing teamwork and communication skills.

Fathers of Biology

Invite children to divide into groups and choose a different scientist to research.

1

2

Provide children with resources to conduct their research, such as books, articles, or websites.

Have each group create a presentation to share with the class about their assigned scientist and their contributions to science.

3

03 Implement icebreakers and team-building activities

Icebreakers and team-building exercises create a positive learning environment that encourages children to work together. These activities teach children to cooperate, respect each other, and feel like they belong, building classroom community.

Create and Test Paper Airplanes

#1

Team-building activities such as having children build and test paper airplanes and study flight are excellent. This practice develops critical thinking, problem-solving, creativity, and teamwork in children. Children must collaborate, communicate, and solve problems to design, make, and test paper airplanes. This helps them value one another's contributions and acknowledge their teammates' skills. Flight science may also teach children about lift and drag and how to apply them to real-world problems. This practice helps improve children's learning and relationships.

Divide children into 3-4 groups.

1

2
Provide paper, scissors, and tape.

Have children design a paper airplane using flight principles.

3

4
Let children fly their paper airplanes and compare designs.

Have children explain their paper airplane's science.

5

6
Encourage children to share and collaborate.

BACKPACK SCIENCES

Scientific Scavenger Hunt

#2

A good team-building activity for children is to send them on a scientific scavenger hunt since it motivates them to cooperate to find and collect scientific materials. In this task, learners must locate and recognize several scientific materials or items using their understanding of science. This encourages cooperation and communication, since they must cooperate to finish the job. The exercise can also be planned so that learners work in small groups, which encourages teamwork and relationships among classmates.

Invite the children to form small teams–for instance, 4-5 children.

1

2

Make a list of science-related activities, such as using a magnet to find something, studying a plant, or classifying various rocks.

Give the children a map of the classroom or school so they can navigate the scavenger hunt.

3

4

Provide various activities for each team to select from, then give them a deadline to finish the assignments.

Encourage children to collaborate, solve problems, and talk to others on their team.

5

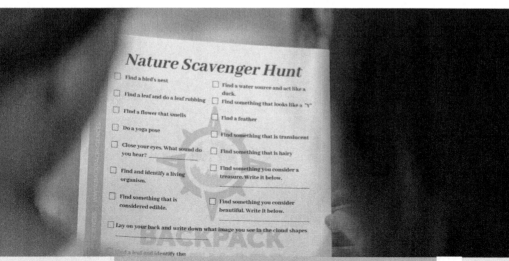

Science-Based Debate

Having children engage in a science-based debate, in which they take on different roles and argue various sides of a scientific topic, can exemplify a positive team-building activity. This type of activity allows children to practice critical thinking and communication skills, as well as work collaboratively in a team setting. As they research and present arguments, children are encouraged to respect and listen to differing opinions and perspectives. Additionally, this activity promotes a sense of ownership and responsibility in their learning, as they can present contribute to the discourse and actively engage with the scientific topic at hand.

Divide the children into two teams and assign each team a side of the debate.

1

2

Either provide or allow the children to research resources and information on the scientific topic to help them prepare their arguments.

Set up the debate format, including opening statements, rebuttals, and closing arguments.

3

4

Encourage the children to use evidence-based arguments and respect each other's opinions.

Have the rest of the class discuss the most compelling arguments.

5

BACKPACK
SCIENCES

STEM Challenge

Having children work together in a "STEM challenge" where they design and build a structure that can withstand an earthquake simulation exemplifies positive team-building. It requires children to collaborate and communicate effectively in order to successfully complete the challenge. By working in teams, children can share ideas and skills, as well as support and encourage each other throughout the design and building process. Additionally, the challenge encourages creativity and problem-solving, as children must consider a range of factors such as materials, structure, and stability when designing their structure. The final testing phase allows for children to see the results of their hard work and encourages them to reflect on their process, learn from mistakes and celebrate successes as a team. Overall, this activity promotes a sense of community and teamwork in the classroom as children work together towards a common goal.

Have the children divide into groups of 3-4.

Provide materials such as paper, tape, straws, and popsicle sticks.

Explain the challenge: the teams must build a structure that can withstand an earthquake simulation, such as shaking the table or dropping a weight.

Set a time limit for the teams to design and build their structures.

Test the structures using the earthquake simulation and observe which structures held up the best.

Encourage the children to reflect on their design process and share their findings with the class.

Group Research Project

Encourage children to conduct a group research project on a scientific topic and present their findings to the class. This is a positive team-building activity as it encourages collaboration, communication, and the sharing of ideas. By working together on a research project, children learn to delegate tasks, brainstorm solutions to problems, and support each other in achieving a common goal. Additionally, presenting their findings allows children to practice their public speaking skills and receive feedback from their peers, while also providing an opportunity to learn from each other's research. Overall, this activity promotes teamwork and encourages children to develop their critical thinking and problem-solving skills in a supportive and collaborative environment.

Invite the children into groups larger than 3.

Invite each group to select a different scientific topic to research.

Provide resources such as books, articles, and websites to help the children with their research.

Set a deadline for the research and have the children work on their presentations.

Encourage the children to collaborate and communicate with their team members.

Invite each group to present their findings and encourage the rest of the class to ask questions and give feedback.

04 Establish clear classroom norms and expectations

Establishing clear classroom norms and expectations is an important aspect of building community among children, as it creates a sense of structure and shared responsibility within the classroom. By establishing clear expectations for behavior and academic performance, children can understand what is expected of them and feel more confident in their ability to succeed.

 Create a Class Norms List

To create a list of class norms together that emphasizes the importance of safety and respect in conducting scientific experiments, an educator can:

Initiate a class discussion on what norms are and why they are important in a classroom setting.

1

2 Ask the children to help brainstorm behaviors that contribute to a positive learning environment, specifically when conducting scientific experiments.

Have children vote on their top choices for class norms and write them on a poster or whiteboard.

3

Ethics Debate

#2

To conduct a class discussion on the ethics of scientific experimentation and have children debate different perspectives, an educator can:

Provide children with a scientific case study that highlights ethical considerations.

Divide children into small groups and assign them different perspectives to argue for or against.

After each group presents, encourage the class to ask questions and provide feedback.

Lab Safety Video

#3

To have children work in groups to create a class lab safety video that emphasizes the importance of safety and responsibility in the laboratory, an educator can:

Provide children with a list of lab safety guidelines and requirements.

Divide children into small groups and have them storyboard and script a video.

Provide children with resources, such as video-editing software or a camera, to create their video.

BACKPACK
SCIENCES

Safety Guidelines

#4

To conduct a scientific inquiry into the effects of different chemical substances and have children reflect on the importance of following safety guidelines in conducting experiments, an educator can:

Provide children with a safe space to conduct experiments, such as a science lab or outdoor area.

1

2

Outline safety guidelines for handling chemicals and provide children with proper protective equipment.

After conducting the experiment, have children reflect on the importance of following safety guidelines and consider what could have gone wrong if guidelines were not followed.

3

Group Contract

#5

To create a group contract that outlines the expectations for group projects, including roles and responsibilities and the importance of communication and collaboration, an educator can:

Divide children into groups and have them work together to create a contract.

1

2

Provide a template or guide for children to use in creating their group contract.

Review contracts with each group and have children sign the contract, demonstrating their agreement to the terms.

3

05 Use group projects and collaborative activities

Building a sense of community among children can be effective through group projects and group activities. Children can learn to communicate, problem-solve, and collaborate effectively by working together to achieve a common goal.

#1 Garden Design

1 Give children a place to design and create their garden—for example, an outside space or a raised garden bed.

2 Invite children to work in teams to design and build a sustainable garden.

3 Outline the scientific concepts involved in gardening, such as soil composition and photosynthesis.

4 Have children choose specific roles and responsibilities to establish clear tasks and duties, such as planning the garden's layout, finding and choosing suitable plants, and caring for the garden.

5 Give children materials to use in creating their gardens, such as soil and gardening tools.

Scientific Research

#2

An educator can have learners complete a group research project on a scientific subject—for example, genetic engineering or biotechnology—and then have them present their findings to the class.

Give them access to materials on the scientific subject, such as books, articles, and videos.

Encourage children to conduct research using a range of sources, including primary and secondary sources.

Children may be divided into smaller groups, with each group choosing a particular study question or area of concentration.

Ask each group to share its results with the group and to learn from one another's work.

#3   Science Fair

Give children instructions for carrying out their own scientific experiments or projects.

Hold a class "science fair" where children present their own scientific experiments or projects.

Hold a "science fair" for the class during which each learner presents their project or experiment.

Have children choose their preferred project. Encourage children to be imaginative and select subjects that fascinate them.

Give them the tools and assistance they need to complete their projects or experiments.

#4   Environmental Condition Investigation

Have students participate in a group investigation into how various environmental conditions, such as temperature and light, affect plant growth, and then have them report their findings to the class.

Describe the scientific theories behind how plants grow and how the environment affects them.

Encourage children to conduct their experiments using scientific approaches, such as gathering data and evaluating outcomes.

Give children a secure environment to do experiments, such as a science lab or a green park.

Separate the children into smaller groups, and give each group a particular environmental aspect to investigate and test.

Ask each group to share its results with the class.

#5   Rube Goldberg

Have children work in groups to construct a "Rube Goldberg machine" that includes scientific ideas like potential energy and kinetic energy.

Children may be divided into teams, with each team being given a certain task or machine part.

Give them the help and direction they need to develop their machines.

Give children the tools to understand the concept, such as materials and examples of Rube Goldberg machines.

Encourage children to use their imaginations and scientific principles when designing their machines.

Set aside time for reflection and sharing

Setting aside time for reflection and sharing can help children feel more connected to one another and to the science learning community. By reflecting on their learning and sharing their ideas and perspectives with one another, children can gain a deeper understanding of the importance of science and their own role within the scientific community.

#1 Science Notebook

Have children keep a science notebook throughout the year to reflect on their learning and personal connections to science.

Introduce the concept of science notebooks and explain their purpose.

Provide children with prompts or guidelines for their notebook entries.

Set aside regular time in class for children to work on their notebooks, such as at the beginning or end of class.

#2 Mistakes Are Great!

Conduct a class discussion on the importance of failure and perseverance in science and have children share examples of times when they overcame challenges in their own scientific endeavors.

Introduce the concepts of failure and perseverance in science and explain why they are important.

1

2 Provide children with examples of famous scientists who faced setbacks in their work.

Encourage children to share their own experiences of overcoming challenges in science.

3

#3 Scientific Slam Poetry

Have children work in pairs to create 'scientific slam' poetry that explores scientific concepts or the process of scientific inquiry.

Introduce the concept of slam poetry and explain how it can be used to explore scientific concepts.

1

2 Provide children with examples of scientific slam poetry.

Pair children up and have them work together to create their own scientific slam poetry.

3

#4 Reflection

Conduct a class reflection on the impact of science on society and have children share their own perspectives on the role of science in their lives and communities.

Introduce the concept of science in society and explain its importance.

1

2 Provide children with prompts or questions to guide their reflections.

Set aside regular time in class for children to share their reflections with the class.

3

BACKPACK SCIENCES

Science Memory Book

#5

Create a "science memory book" at the end of the year that includes personal reflections and highlights from the year in science.

Provide children with a template or guide for creating their memory book.

Provide children with time in class to work on their memory books and encourage them to share their books with the class at the end of the year.

1

2

3

Encourage children to reflect on their favorite moments from the year in science and what they have learned.

07 Engage families and the broader community in student learning

Engaging families and the broader community in student learning is important because it helps to create a sense of shared responsibility for student success and fosters a sense of community and support for the children. When families and community members are involved, children can feel more connected to their families and communities and see the real-world applications of science in their everyday lives.

To implement these activities in a learning environment setting, educators can follow these procedures:

Science Family Night

Help children and their families engage in science activities together.

Plan and organize the event well in advance and ensure that all necessary materials are available.

1

2

Send out invitations to families well in advance and provide clear instructions on what to expect.

Plan activities that are engaging and interactive, and that can be completed by families together.

3

4

Set up stations around the classroom or school that focus on different scientific concepts or experiments, and have families rotate through each station.

Science Museum

Create a class "science museum" that showcases the different scientific experiments and projects completed throughout the year and invite families and other classes to visit.

Set aside a designated area in the classroom or school to display the science museum.

1

2

Assign different groups of children to create exhibits that showcase their scientific learning from the year.

Provide clear guidelines for creating the exhibits, such as using clear labels and including interactive components.

3

4

Invite families and other classes to visit the science museum during designated times, and have children act as guides to explain their exhibits.

BACKPACK SCIENCES

#3 Scientist Invitation

Invite scientists or STEM professionals from the community to visit the classroom and speak about their experiences and careers.

1 Identify local scientists or STEM professionals who would be willing to visit the classroom.

2 Coordinate with the visitors to ensure that their visit aligns with the curriculum and learning objectives.

3 Prepare children in advance by discussing the visitors' backgrounds and areas of expertise.

4 Provide children with questions to ask the visitors, and encourage them to think critically about the information presented.

#4 Community Service

Have students work in groups to create a community service project that incorporates scientific principles, such as creating a composting program or planting trees in the community.

1 Brainstorm potential community service projects that incorporate scientific principles, and choose one that is feasible and appropriate for the children's ages.

2 Divide children into groups and assign specific roles and responsibilities for each group member.

3 Provide clear guidelines and expectations for the project, including deadlines and necessary materials.

4 Invite families and community members to participate in the project, and provide opportunities for children to present their work to the community.

Scientific Storytelling

#5

Conduct a "scientific storytelling" event in which students share personal stories about their experiences with science, and invite families and community members to attend.

Provide children with clear guidelines for creating their scientific stories, such as including a beginning, middle, and end, and focusing on a particular scientific concept or experience.

Invite families and community members to attend the event, and provide refreshments and opportunities for discussion following the storytelling.

1

2

3

Provide opportunities for children to practice sharing their stories in small groups or with partners.

08 Encourage leadership and decision-making

To foster a sense of community, the educator can encourage science leadership and decision-making opportunities. By giving kids the ability to take charge of their learning, make decisions, and take responsibility for their activities, they build a sense of ownership and involvement in their learning environment. When given the chance to lead and make decisions, children learn critical thinking, problem-solving, and communication skills that will assist them throughout their lives. Fostering science leadership and decision-making can create a good, supportive learning atmosphere where children feel empowered and engaged.

BACKPACK SCIENCES

Citizen Science Project

Citizen science project may foster leadership and decision-making. These projects let children solve real-world problems and get involved in their community, giving them the chance to make decisions and own their work. Working with local groups on a Citizen Science project gives children the chance to gather and evaluate data, cooperate with others, and build leadership skills that will help them thrive in school and work. Children can learn science and help their communities by participating in Citizen Science programs.

Find a local Citizen Science project that fits your curriculum or students' interests. Examples include water quality, bird migration, and local flora and fauna.
https://www.citizenscience.gov/#

Contact the organization to explore partnership and student data collection techniques.

Create a project plan with goals, methodology, and data-gathering instruments. Work with the organization to ensure the project meets their needs.

Introduce citizen science to the students and explain its goals. Discuss data collection methods, tools, and safety.

Use agreed-upon procedures and instruments. Make sure youngsters are safely recording accurate data.

Analyze project data with the children. Discuss data trends and outliers.

Work with children to prepare a report or presentation of your findings. Present the report to the collaborating organization and the community.

Discuss what children learned, what they liked or didn't like, and how they may apply their skills and knowledge to future projects.

Science Journalism

Engaging children in a science journalism project can promote leadership and decision-making skills while also improving science communication abilities. By participating in this project, children can take on a leadership role in selecting and researching science topics of interest, making decisions on how to best communicate complex scientific concepts, and presenting their findings to their peers and community. In addition, this project can help foster critical thinking, teamwork, and creativity as children work together to craft compelling and informative science articles.

1 Explain to the children they will be creating science articles for their school newspaper or science magazine. Discuss the importance of science communication and how their articles can inform and educate the wider community.

2 Allow children to choose their own topics, or assign topics related to current events or recent scientific discoveries. Encourage children to choose topics that they are passionate about.

3 Provide children with resources such as scientific journals, news articles, and online databases to conduct their research. Encourage students to use a variety of sources to ensure their articles are accurate and well-informed.

4 Provide children with a template or structure to follow when writing their articles. This could include an introduction, background information, key findings, and a conclusion. Encourage students to write in a clear and concise style, avoiding jargon and technical terms that may be difficult for readers to understand.

5 Have children peer review each other's articles. This will help them develop critical thinking skills and learn how to provide constructive feedback.

6 Once children have received feedback from their peers, they should edit and revise their articles accordingly.

7 Publish the articles in the school newspaper or science magazine. Encourage children to share their articles on social media or other platforms to reach a wider audience.

8 After the project is complete, encourage children to reflect on their experience. Discuss what they learned about science communication, writing, and research. Encourage them to identify areas for improvement and to think about how they can apply their skills in the future.

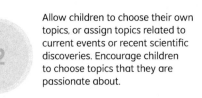

BACKPACK SCIENCES

Science Olympiad

Encouraging children to participate in science competitions such as Science Olympiad can be an excellent way to promote leadership and decision-making skills. Science Olympiad, a nationwide nonprofit, improves science education via hands-on, interactive, and inquiry-based competitions. These competitions motivate children to work together to tackle difficult challenges and compete in scientific activities. Science Olympiad challenges children to think critically, make choices, and take responsibility. This form of competition gives children the chance to learn leadership, teamwork, and complicated problem-solving.

1 Research Science Olympiad. Examine its rules, events, and format.

2 Invite interested children to form a Science Olympiad team (15 children maximum).

3 The team can then choose their events. Science Olympiad has 23 events in biology, chemistry, physics, earth science, and engineering.

4 Give team members roles depending on their abilities and interests. For example, one child may be skilled in biology and study for the biology event, while another might prefer engineering and build the team's engineering entry.

5 The squad can practice for events. They can research, test, and build prototypes.

6 Help the team use resources such astextbooks, scientific publications, lab equipment, and computers.

7 Support the team as they compete in their events in regional, state, or national Science Olympiad competitions.

8 Celebrate the team's successes, whether they won or just participated and learned. The team might inspire learners to join Science Olympiad by sharing their experiences.

Science Olympiad
Resources

Science Olympiad:

The official Science Olympiad's website details events, rules, and resources for coaches and students. The webpage also explains how to form a Science Olympiad team.
https://www.soinc.org/

National Science Olympiad:

The competition, past winners, and coach and resources for participants are on the website. The website offers event-specific sample questions and study questions.
https://www.soinc.org/about-national-science-olympiad

Science Olympiad Foundation:

This nonprofit organization supports science education through competitions and Olympiads. The website covers science Olympiads and tournaments, notably the Science Olympiad.
https://www.sofworld.org/science-olympiad-foundation

YouTube:

Several channels offer Science Olympiad tools and tutorials. Science Olympiad TV, Tips, and Mentor are popular channels.

Scioly.org:

Scioly.org is a community-driven website with Science Olympiad resources, forums, and study tools. Students and coaches can collaborate on the website.
https://scioly.org/

Local science museums and organizations may offer resources, workshops, and training programs for Science Olympiad participants. It's worth reaching out to these organizations to see if they offer any support or resources for Science Olympiad teams.

BACKPACK
SCIENCES

Service Learning

#4

Having children participate in service learning, which involves engaging children in community service projects that are tied to their academic learning, can help build community among children while also promoting scientific knowledge and skills. Service learning projects can be science-related, such as cleaning up a local park or beach, conducting water quality testing, or creating a community garden to study plant growth and ecology. These projects provide opportunities for children to engage with their community, apply scientific concepts in real-world contexts, and develop leadership and teamwork skills.

1 Identify a science-related service project that aligns with the curriculum and the students' interests. This could include cleaning up a local park or beach, planting trees, building birdhouses, or other projects that benefit the environment or community.

2 Reach out to local organizations or groups that are involved in the service project to establish partnerships and coordinate logistics. This could include local parks and recreation departments, environmental groups, or other community organizations.

3 Present the service project to the children and explain how it connects to the science curriculum. Emphasize the importance of community involvement and service learning.

4 Work with children to plan and prepare for the service project. This could include researching the environmental issue at hand, designing and building tools or equipment needed for the project, and determining the roles and responsibilities of each team member.

5 Coordinate with the community partners to execute the service project. Ensure that children have the necessary tools and equipment and that safety protocols are in place. Encourage children to work collaboratively and take on leadership roles throughout the project.

6 After the service project is complete, encourage children to reflect on their experience. This could include discussing the impact of the project on the community, reflecting on what they learned about the environment and science, and identifying areas for future growth and improvement.

7 Finally, encourage children to share their experience and knowledge with others. This could include presenting the project to the class or the school, creating a blog or social media post about the experience, or sharing the project with the local community.

Experimental Design Challenge

#5

Help children develop leadership and decision-making skills while solving real-world problems. Children can learn to solve complex problems through experimentation. This exercise empowers children to learn and collaborate. Working on real-world challenges can also motivate and engage children in science.

1 Introduce the real-world problem—for example, reducing plastic waste or increasing air quality. Discuss the value of scientific experimentation in solving these issues.

2 Tell children that they will collaborate in teams to plan and conduct an experiment to tackle a real-world problem. Emphasize that they will design, collect, analyze, and draw conclusions from the experiment.

3 Have children brainstorm experiments to solve the situation. Encourage them to be creative and consider of cost, feasibility, and impact.

4 After brainstorming, have children work in teams to develop experimental designs. Emphasize clear procedures and changeable control.

5 After designing experiments, give children the materials they need. Examples include lab equipment, measurement devices, and data collection materials.

6 Give children time to experiment. Answer inquiries and offer advice.

7 After the experiments, have children examine the data. Encourage graphs, charts, and statistical analysis.

8 After analyzing facts, have children draw conclusions. Encourage students to critically evaluate their discoveries and apply them to real-world issues.

9 Have children present to the class or a panel of parents. Encourage them to use PowerPoint or films to improve their presentations.

10 Have children reflect on the presentations. Discuss what they learned, what they would change, and how to enhance their designs. Encourage them to apply their new talents to other real-world issues.

09 Foster a growth mindset and celebrate achievements

Fostering a growth mindset and celebrating achievements are essential components of building a positive and inclusive classroom environment in which children can thrive. Science education, where scientific inquiry and experimentation can be difficult, requires this. Educators may help children develop resilience, perseverance, and curiosity for science and life by teaching a growth mindset and applauding successes. Sharing and celebrating successes with peers strengthens the learning environment. Fostering a growth mindset and rewarding successes can improve children's learning results and build a supportive school culture that benefits everyone.

Science Vision Board Activity

The scientific vision board activity allows children to think about and express their science ambitions. This fosters a supportive environment where children can help each other achieve their goals.

Explain the exercise and supply magazines, stickers, and paper. Show children examples of vision boards.

1

2

Have students make a vision board with science goals. What science topics do they want to learn this year? What science topics are they interested in or enjoy? What aspects of nature make them happy?

Allow children to present their vision boards.

3

Perseverance and Growth Mindset Discussion

Having a discussion on perseverance and growth mindset allows children to share their own experiences of overcoming challenges in science and inspire each other to adopt a growth mindset.

Growth mindset means believing that you can learn and improve through hard work and effort. In science, having a growth mindset means understanding that making mistakes and experiencing failure is a normal part of the learning process. Scientists who have a growth mindset believe that they can overcome challenges and improve their skills with practice and determination.

In other words, perseverance and growth mindset in science are all about never giving up and believing that you can learn and improve with hard work and practice. So, if you want to be a great scientist, remember to keep trying and never give up, even when things get tough!

Introduce the topics of perseverance and growth mindset in science. Perseverance means not giving up, even when something is difficult or challenging. In science, it's important to have perseverance because scientific discoveries often require a lot of hard work and patience. Scientists have to keep trying, even if their experiments don't work the first time.

1

2

Ask children to share their own experiences of overcoming challenges in science and how they persevered through difficult tasks.

Facilitate a discussion on the benefits of a growth mindset and the importance of perseverance in science.

3

BACKPACK
SCIENCES

#3   Science Achievement Wall and Celebration Day

The Science Achievement Wall and Celebration Day recognize the children's individual and collective scientific achievements, creating a sense of pride and belonging in the classroom community.

Create a Science Achievement Wall in the classroom that highlights the accomplishments of individual students and the class as a whole.

Brainstorm with the children what criteria a child should possess or accomplish to be on the wall.

Allow children to nominate or suggest their peers.

Conduct a Science Celebration Day in which students share their own scientific projects and accomplishments and engage in fun scientific activities together.

Scientific Gratitude Card Activity

#4

The scientific gratitude card activity allows children to show their appreciation for scientific concepts or figures and promotes gratitude and positivity in the classroom.

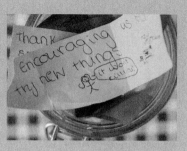

1

Introduce the activity. Scientific ideas and figures help us understand the world and how it works to improve our lives.

For example, photosynthesis helps us understand how plants create food and how vital they are to our survival. Learning about butterfly life cycles helps us appreciate nature and how organisms evolve through time.

We also admire scientists who have achieved life-changing discoveries or creations. We admire Marie Curie, who discovered radium and polonium and advanced radiation research. We admire Thomas Edison, who invented the light bulb and made electricity more accessible worldwide.

Thus, scientific concepts and figures help us understand the world and improve our lives. Science and its contributors can help us appreciate its wonders.

Give examples such as gravity, the periodic table, Marie Curie, etc.

2

3

Show children how to make a scientific gratitude card. Invite them to decorate, draw, and write words of gratitude.

Encourage children to make scientific gratitude cards alone or with partners. Encourage them to design and write thoughtfully.

4

5

After finishing the cards, have children trade. Students should read the remarks and applaud each other.

Discuss how expressing thankfulness in science might help learners develop a growth attitude. Encourage children to discuss the exercise.

6

7

Give children a bulletin board or gratitude jar to exhibit their scientific gratitude notes. Have children add to the display all year.

Growth Mindset Journals

#5

Children can keep growth mindset journals in which they reflect on their progress and growth in science and set goals for future learning. At first, sharing encourages a sense of vulnerability, but over time, the children will consider this a space where failures are accepted and supported.

Introduce the concept of a growth mindset journal to children and explain that they will use it to reflect on their progress and set goals.

1

2

Provide children with journals or notebooks to use for their reflections.

Encourage children to write about challenges they've faced in science, and how they've persevered and grown from those challenges.

3

4

Provide time for students to reflect on their progress and set goals for future learning. Provide positive feedback and recognition for their efforts.

Each week, invite children to share a failure and what they learned from the process.

5

10 Incorporate technology and multimedia

Incorporating technology and multimedia can build community among children by providing opportunities for them to work collaboratively on projects and share their creations with one another. This can also help children develop digital literacy skills and engage with scientific concepts in a fun and interactive way.

Science Podcast

Creating a science podcast is a great example of incorporating technology and multimedia to build community among children. Through this activity, children can explore and research different scientific concepts, interview scientists and STEM professionals, and share their knowledge with their peers. By working collaboratively to create a podcast, children can learn from one another and develop important communication and teamwork skills. Additionally, sharing their podcast with the larger school community can help to foster a sense of pride and achievement in their work. Overall, creating a science podcast is a fun and engaging way to encourage children to embrace technology and multimedia while building a sense of community in the classroom.

1 Introduce the concept of a science podcast to the children and explain its purpose, which is to share scientific knowledge and inspire others to learn more about science.

2 Discuss different science topics that can be explored through the podcast and encourage the children to brainstorm ideas for their own podcast episode.

3 Invite the children to divide into small groups and allow each group to choose a topic of interest to research and discuss in their podcast episode.

4 Provide the children with access to resources such as books, articles, and online sources to help them with their research.

5 Have the children work collaboratively to create a script for their podcast episode, including an introduction, discussion of their topic, and a conclusion. An example or a template is optional.

6 Assist the children with recording and editing their podcast using available technology, such as a tablet, computer or phone.

7 Encourage the children to share their podcast with the larger school community, either through a classroom presentation or by uploading it to the school website.

Experimental Design Challenge

#2

An experimental design challenge can help children develop leadership and decision-making skills while solving real-world problems. Children learn to solve complex problems through experimentation. This exercise empowers children to learn and collaborate. Working on real-world challenges can also motivate and engage children in science.

1 Choose a scientific topic that aligns with your curriculum and your children's interests.

2 Explain to the children the purpose and goals of creating a science documentary project and how it will enhance their understanding of the topic.

3 Invite the children to divide into small groups and assign them specific roles, such as researcher, scriptwriter, producer, editor, and presenter.

4 Provide resources such as books, articles, and online databases for students to conduct their research.

5 Set a timeline for each phase of the project, including research, scriptwriting, filming, editing, and final presentation.

6 Monitor and provide guidance to each group throughout the project, including checking their scripts and footage, and providing feedback on their progress.

7 Allow time for children to practice and rehearse their presentations.

8 Encourage marketing features to promote their documentary—for example,example, a movie poster, social media post, or video trailer.

9 Host a screening event where each group can present their documentary to the class or school community.

10 Facilitate a Q&A session after each screening where children can ask questions and provide feedback to their peers.

BACKPACK SCIENCES

Science Infographic

Creating a scientific infographic with children is a fun way to introduce technology and multimedia into the classroom and develop community. Children can communicate scientific topics creatively and interactively by drawing them. Infographics help learners think critically about their subject matter and how to convey it graphically. This activity promotes digital literacy and class togetherness.

1

Explain infographics and show science examples. An infographic is an entertaining, easy-to-understand visual. It's like a story or explanation in an image.

Pictures, graphs, and other visuals help you understand concepts. Fun colors, huge numbers, or hilarious pictures make them more engaging.

If you wanted to learn about ocean animals, you might locate an infographic with pictures and fun facts. If you wanted to know how many people spoke different languages, you could find an infographic with a big colorful chart.

Infographics are fun and quick ways to learn new things!

Discuss data visualization, text, and images in infographics.
Show examples.

Give children research resources—for example, scientific publications or data sets.

Allow time for graphical peer review and comments.

Assign a scientific theme or let children choose their infographic topic.

Use Canva, Piktochart, or Google Slides to have children create infographics in pairs or small groups.

Have children explain their infographics to the class.

Virtual Science Fair

#4

Creating a virtual science fair allows children to use technology to display their scientific experiments or projects and share them with others, which can develop community and teamwork. Children can ask questions, give feedback, and learn from each other in a virtual scientific fair.

1 Educator preparation: Coordinate a day for a virtual science fair in which children from all over the world can share their projects.

2 Explain a virtual science fair to children.

A virtual science fair allows children to display their science projects and experiments online. Children can share their science fair projects online instead of in person using videos, photos, and other digital tools.

Like in a regular science fair, children can get feedback on their projects and learn from others. They can also see what other children their age are interested in and how they chose to further explore topics.

Virtual science fairs allow students from around the world to participate. Students can interact with more individuals and obtain input from those with diverse viewpoints.

A virtual science fair is a great method for children to share their passion for science and learn more about it!

Set a theme or topic for the virtual science fair and provide guidelines for projects—for example, due dates and check-ins.

3

4 Have children work individually or in groups to plan and conduct their experiments or research projects.

Ask children to use technology to create digital presentations of their projects, such as slideshows, videos, or interactive websites.

5

6 Set up a platform for children to share their projects, such as a class website or learning management system.

Encourage children to view and interact with each other's projects and provide opportunities for them to ask questions and provide feedback to their peers.

7

BACKPACK SCIENCES

Science Animation

#5

Creating a science animation can foster collaboration and communication among children as they work together to visually represent scientific concepts or experiments. Children can produce entertaining and educational animations using animation software or apps and share them with the class or online. This project helps children improve digital literacy and creativity.

Introduce the scientific concept the students will animate. A lecture, video, or materials for children to research could do this.

1

2

Provide access to animation software. Provide guidance and make sure all learners understand how to use the programs.

Have children brainstorm animation concepts in pairs or small groups. They should decide what scientific principles to explain and how to use animation to communicate them.

3

4

Make an animation storyboard. Scenes, people, and conversations should be sketched out.

Start animating with the tools. The educator should answer technical inquiries and provide direction.

5

6

After animating, children should review and revise. The educator can assess animation quality and scientific accuracy.

Finally, children can present their animations. The educator can screen student animations or create a digital platform for them to upload the projects. Children can critique and assist each other's animations.

7

COMMON CHALLENGES EDUCATORS FACE IN BUILDING COMMUNITY

#1

I DON'T KNOW IF THIS IS GOING TO WORK. THE CHILDREN AND I DON'T REALLY LIKE CHANGE.

Some children or educators may resist new community-building strategies, especially if they are used to more traditional teaching methods.

Solution:

Educators can work to build support for new strategies by explaining the benefits and getting buy-in from administration, children or co-educators. Administration can also offer training or professional development opportunities to help educators feel more comfortable with new approaches.

#2

THE CHILDREN DON'T SEEM TO BE INTERESTED. THEY'RE NOT ENGAGED.

It can be challenging to to engage all children in community-building activities, especially if they are not interested in science or do not feel confident in their abilities.

Solution:

To overcome this challenge, educators can work to create activities that appeal to a wide range of interests and abilities, offer incentives or rewards for participation, and provide opportunities for students to work in teams or collaborate with others.

#3

I DON'T HAVE TIME TO BUILD COMMUNITY. I NEED TO TEACH THE CHILDREN THE BASICS OF MATH, LANGUAGE AND READING.

Time constraints: Building a community may seem like an extra burden on top of teaching the basics of math, language, and reading. However, building a community can actually support and enhance learning, especially in science. It provides opportunities for children to collaborate, develop leadership skills, and apply critical thinking. Furthermore, it can improve engagement, motivation, and overall well-being.

Solution:

Start small. There are many simple strategies and activities that can be integrated into daily routines, such as morning meetings, cooperative learning activities, and service learning projects. By starting small and building gradually, educators and parents can create a sense of community without overwhelming their already busy schedules.

BACKPACK SCIENCES

5 QUICK FIXES

Five small, doable suggestions that can lead to quick results for educators or homeschooling parents looking to incorporate community building into their science lessons

1. CELEBRATE SUCCESS

Take the time to celebrate individual and group successes in class, such as completing a difficult experiment or project.

2. USE ICEBREAKERS

Start class with icebreaker activities that allow students to get to know each other and feel comfortable sharing their ideas.

3. INCORPORATE PEER TEACHING

Allow students to teach each other by assigning them topics to research and present to the class.

4. CONDUCT COMMUNITY OUTREACH

Encourage students to participate in science-related community service projects, such as cleaning up a local park or stream.

5. USE INQUIRY-BASED LEARNING

Encourage students to ask questions and explore their interests through inquiry-based learning activities that allow for creativity and critical thinking.

SUMMARY

Discovery, experimentation, and innovation contribute to fascination with science. However, many children find science intimidating. As educators, it is our responsibility to find ways to engage children in science so that they become lifelong learners and develop a passion for the subject. Community building can help accomplish this.

Community building fosters a sense of belonging, respect, and support. Community building in science education helps children feel comfortable investigating and experimenting with scientific concepts. Creating a community in the learning environment can help children appreciate science and view themselves as capable learners.

Group science projects can help develop community. Collaborative learning assists children to develop communication, problem-solving, and critical thinking. Group projects may include science fairs, class experiments, or service learning projects. These projects enable children to collaborate, share ideas, and learn together.

Another community-building method is citizen science projects. Citizen science lets children solve real-world problems and examine data. These projects give children opportunities to discover how science applies to their lives and take ownership of their learning. Citizen science projects also allow children to work with scientists, which allows behind-the-scenes opportunities and may motivate children to become scientists.

Additionally, science journalism projects can engage the community in science education. By having children produce science pieces for their school newspaper or a science magazine, educators can encourage learners to strengthen their writing abilities, interviewing skills, and technology while also learning about scientific subjects. These projects also let children show off their work, which can boost their confidence.

BACKPACK
SCIENCES

Science Olympiad is another great way to establish community. Science competitions help children work on tough projects, develop problem-solving skills, and compete with students from around the world. Children can network with professionals while earning recognition at these tournaments.

Science education service learning initiatives develop community, too. Service learning projects—for example, cleaning up a park or beach—can help children develop responsibility and civic participation. These projects can also provide children a chance to collaborate with their neighbors, which can foster community.

Encourage learners to participate in experimental design challenges to establish a science education community. Educators can improve students' critical thinking and appreciation of science by having them plan and execute experiments to tackle real-world problems. These challenges allow children to solve complex problems creatively, which can boost their confidence as learners.

Technology and multimedia in science teaching can nurture community. Educators may assist children to develop creativity and technology skills by having them produce science podcasts, movies, infographics, animations, and virtual science fairs. These projects also allow children to show off their work, which can boost their confidence.

However, educators may face challenges in implementing these strategies, such as time constraints, lack of resources, and student apathy. These challenges can be addressed by incorporating small, doable suggestions like using interactive activities, incorporating local and relevant science topics, and setting achievable goals.

Community building can help children embrace science by creating a good learning environment. Through hands-on activities and collaborative projects, educators may help children become curious, creative, and confident lifelong learners who love science.

CONCLUSION

In conclusion, *Teaching Montessori Science : 9 Practical Strategies to Engage Children in Hands-On STEAM Activities* underscores the importance of hands-on, interactive, and experiential learning in science education. Throughout this book, we have explored innovative approaches to science instruction that prioritize student engagement, cultural sensitivity, interdisciplinary connections, and real-world relevance. Our journey has been guided by the belief that by empowering educators with practical strategies and insights, we can inspire a lifelong love of science in children, fostering curiosity, creativity, and critical thinking skills that extend far beyond the classroom walls.

Throughout "*Teaching Montessori Science : 9 Practical Strategies to Engage Children in Hands-On STEAM Activities*," we have looked into the crucial role of hands-on, interactive, and experiential learning activities in science education.

Chapter 1 emphasizes the transformative impact of such activities on science knowledge, engagement, and creativity in children, highlighting the significance of incorporating physical exercises, field trips, and experiential learning to foster a lasting passion for science.

Moving forward, Chapter 2 advocates for culturally sensitive teaching methods to effectively engage children in science education, stressing the importance of inclusivity, diversity, and cultural relevance in creating an empowering learning environment.

In Chapter 3, we explore the value of interdisciplinary approaches in making science education more relevant and engaging, showcasing examples of interdisciplinary projects that enhance learning outcomes.

Next, Chapter 4 discusses the power of storytelling in making science concepts relatable and captivating for children, offering practical strategies to incorporate storytelling into science instruction.

Chapter 5 emphasizes the benefits of offering autonomy and choice to children in their science education, providing insights into how choice can increase engagement and motivation.

Meanwhile, Chapter 6 delves into the role of real-world issues and problem-solving in science education, showcasing strategies to integrate real-world problems into science lessons effectively.

In Chapter 7, we explore the potential of technology to enhance hands-on science exploration and engagement, emphasizing its role as a tool to support learning and foster a love of science.

Moving on, Chapter 8 highlights the importance of observation in scientific inquiry and Montessori education, offering examples of how educators can help children develop observation skills.

Finally, Chapter 9 discusses the significance of community building in science education, providing strategies to create a supportive and inclusive learning environment through collaborative projects, citizen science initiatives, and science competitions.

Together, these chapters offer a comprehensive guide to empowering educators with practical strategies and insights to inspire a lifelong love of science in children, fostering curiosity, creativity, and critical thinking skills.

AFTERWORD

Reflecting on the journey of writing this book, I am struck by the transformation that occurred from its conception to completion. Initially envisioned as a shorter, quick read, excitement and passion ignited within me as I went into the process of brainstorming the myriad strategies to share. Each idea sparked a new wave of enthusiasm, a desire to provide readers with tangible, actionable examples that could be applied immediately in their learning environments.

As the pages unfolded, it became evident that this book was not merely about imparting knowledge but about instilling a sense of empowerment. I wanted readers to feel reassured, knowing that they were not alone in their struggles and that with the practical strategies, ideas, and support offered within these pages, teaching science could indeed be both effortless and enjoyable. This philosophy resonates deeply with my work at Backpack Sciences, where I strive to equip educators with the tools and confidence they need to cultivate a lifelong love of science in their students. Each chapter is a testament to the belief that by embracing creativity, innovation, and community, we can transform the landscape of science education, one educator at a time.

As we draw to a close on this transformative journey through Montessori science education, I extend a heartfelt invitation for you to take action and apply the invaluable insights gained from this book. With practical strategies and real-world examples at your disposal, you hold the key to revolutionizing your approach to teaching science and fostering profound connections with your students.

Consider becoming a member of Backpack Sciences, where science lessons are thoughtfully curated into units, each complemented by a diverse array of multi-disciplinary follow-up activities. As a special offer, I invite you to explore a free sample of our Skeletal Lesson plan from the Human Biology unit, available at https://www.backpacksciences.com/skeletal.

By joining our vibrant community, you'll not only gain access to a wealth of resources and ongoing support but also connect with like-minded educators committed to reshaping the landscape of science education. Together, let's ignite a passion for science in the hearts and minds of our students, empowering them to embark on a journey of exploration, discovery, and innovation.

Visit www.backpacksciences.com today and seize the opportunity to transform your classroom into a hub of curiosity, creativity, and scientific inquiry.

References

Chapter

01

Aguiar Jr, A. S., Castro, A. A., Moreira, E. L., Glaser, V., Santos, A. R. S., Tasca, C. I., Latini, A., Prediger, R. D. S. (2011). Short bouts of mild-intensity physical exercise improve spatial learning and memory in aging rats: Involvement of hippocampal plasticity via AKT, CREB and BDNF signaling. Mechanisms of Ageing and Development, 132(11-12), 560-567.

Bartos, S., Lederman, N. G., Bartos, J., & Oliver, J. (2019). Supporting early science learning through hands-on experiences: The role of inquiry-based science learning. Early Childhood Education Journal, 47(1), 35-42.

Chikwa, G., Fraser, S. P., & Mumba, F. (2019). Enhancing the quality of primary science teaching and learning in Zambia through science-related field trips and visits from scientists. African Journal of Research in Mathematics, Science and Technology Education, 23(1), 17-27.

Ding, Q., Ying, Z., & Gómez-Pinilla, F. (2011). Exercise influences hippocampal plasticity by modulating brain-derived neurotrophic factor processing. Neuroscience, 192.

Driver, R., Asoko, H., Leach, J., Mortimer, E., & Scott, P. (1994). Constructing scientific knowledge in the classroom. Educational Researcher, 23(7), 5-12

Eshach, H., & Fried, M. N. (2019). Hands-on and experiential activities in science education: A systematic review of literature. International Journal of Science Education, 41(18), 2662-2688.

É.W. Griffin, S. M., Carole Foley, S. A. Warmington, S. M. O'Mara & Á. M. Kelly (2011). Aerobic exercise improves hippocampal function and increases BDNF in the serum of young adult males. Physiology & Behavior, 104, 5, 934-941.

Falk, J. H., & Dierking, L. D. (2000). Learning from museums: Visitor experiences and the making of meaning. Walnut Creek, CA: AltaMira Press.

Ferlazzo, Larry. (2020). Eight Ways to Use Movement in Teaching & Learning. EdWeek blog. July 24, 2020.

Jeong, H., Park, J., Lim, J., & Kwon, J. (2018). The effects of group discussion and peer teaching on science learning in elementary school. International Journal of STEM Education, 5(1), 1-9.

Mehta, R. K., Shortz, A. E., Benden, M.E. (2015). Standing Up for Learning: A Pilot Investigation on the Neurocognitive Benefits of Stand-Biased School Desks. International Journal of Environmental Research and Public Health 13, 1.

Rivet, A. E., & Krajcik, J. S. (2004). Achieving standards in urban systemic reform: An example of a sixth grade project-based science curriculum. Journal of Research in Science Teaching, 41(7), 669-692.

Stevens, R. J., Olivares-Donoso, R., & Williams, T. (2008). The effects of project-based science instruction on urban, low-achieving sixth-grade students. Journal of Research in Science Teaching, 45(8), 931-954.

Chapter 02

Akerson, V. L., Carter, I. S., & Park Rogers, M. A. (2016). Incorporating nature of science and scientific inquiry into STEM education. In STEM education (pp. 25-43). Springer.

Bell, P., Lewenstein, B., Shouse, A. W., & Feder, M. A. (2009). Learning science in informal environments: People, places, and pursuits. National Academies Press.

Davis, K. E., & Rimm, E. B. (2018). What students say about the climate for diversity in a medical school. Teaching and Learning in Medicine, 30(1), 1-9.

El-Hani, C. N., Mortimer, E. F., & Zanchettin, J. L. (2010). Meaningful learning in biology and environmental education: A cultural-historical approach. International Journal of Environmental & Science Education, 5(4), 467-481.

Gay, G. (2010). Culturally responsive teaching: Theory, research, and practice (2nd ed.). Teachers College Press.

Kloser, M. (2014). Using science concepts and examples that are relevant to the students' lived experiences: Moving beyond "knowing about" science to making sense of our world through science. Journal of Science Teacher Education, 25(4), 431-449.

Lee, O., Quinn, H., & Valdes, G. (2013). Science and language for English language learners in relation to Next Generation Science Standards and with implications for Common Core State Standards for English Language Arts and Mathematics. Educational Researcher, 42(4), 223-233.

Osborne, J. (2010). Arguing to learn in science: The role of collaborative, critical discourse. Science, 328(5977), 463-466.

Rath, K. (2019). Developing Cultural Competence in Montessori Education. Montessori Life, 31(4), 32-37.

Rosenberg, J. E., & O'Callaghan, E. M. (2019). Beyond gender: Montessori education as an inclusive and responsive environment. Montessori Life, 31(4), 38-43.

Chapter 03

Kiili, K., Perttula, A., Lindstedt, A., Arnab, S., & Suominen, M. (2019). Student engagement and academic performance in science education: A study of the impact of digital game-based learning. British Journal of Educational Technology, 50(2), 566-580. https://doi.org/10.1111/bjet.12693

Linnenbrink-Garcia, L., Durik, A. M., & Conley, A. M. (2017). An interdisciplinary approach to exploring motivation and learning. Journal of Educational Psychology, 109(6), 778-793.

Rathunde, K., & Csikszentmihalyi, M. (2006). Montessori education and optimal experience. Journal of Montessori Research, 2(1), 19-38.

Wright, T., DeBacker, T. K., Lichtenberg, J. W., & Wagner, A. Z. (2020). An integrated STEM approach to elementary mathematics instruction: Exploring effects on engagement and achievement. Journal of Research in Science Teaching, 57(10), 1604-1631. https://doi.org/10.1002/tea.21611

Chapter 04

International Journal of Early Childhood Special Education, 11(2), 100-113. doi: 10.20489/intjecse.594822

Schmitt, M. C., & Schatz, J. (2018). Engaging elementary school students with a story: The influence of storytelling on comprehension and motivation to learn. Journal of Educational Psychology, 110(7), 929-940. doi: 10.1037/edu0000256

Vellutino, A., & Scanlon, D. (2019). The power of storytelling in promoting language development, literacy skills, and social-emotional development in young children.

Chapter 05

Anderson, D., Osguthorpe, R.T., Kimball, R., and Farris, S.P. (2015). The effects of student choice on motivation and learning. Educational Technology Research and Development, 63(2), 259-276.

Johnson, L. (2020). Personalized Learning: What It Is, Why It Matters, and How to Implement It. EdTech. https://www.edtechmagazine.com/k12/article/2020/05/personalized-learning-what-it-why-it-matters-and-how-implement-it-perfcon

Lillard, A. S. (2005). Montessori: The science behind the genius. Oxford University Press.

Lynch, R., and Dembo, M. (2004). The relationship between self-regulation and online learning in a blended learning context. The International Review of Research in Open and Distributed Learning, 5(2), 1-16.

McComas, W.F., and Abraham, L. (2004). The effect of student-generated diagrams versus student-generated summaries on conceptual understanding of science concepts in middle school. Journal of Research in Science Teaching, 41(7), 715-730.

Pekrun, R., and Elliot, A.J. (2009). Achievements and emotions. New York: Routledge.

Plucker, J.A., Beghetto, R.A., and Dow, G.T. (2017). Why isn't creativity more important to educational psychologists? Potentials, pitfalls, and future directions in creativity research. Educational Psychologist, 52(3), 168-186.

Riggan, M., & Mosley, P. (2017). Increasing Student Engagement and Motivation: From Time on Task to Homework. In Teacher Education and Professional Development (pp. 201-221). IGI Global.

Chapter 06

Babic, M., Sherry, A., Mocilnikar, A., & Brenk Sinicki, M. (2018). Promoting sustainability through science education: a review of the literature. Environmental Education Research, 24(4), 527-545.

Chiappetta, E. L., & Koballa Jr, T. R. (2016). Science instruction in the middle and secondary schools. Pearson.

Crawford, B. A. (2014). Next generation science standards: a vision for K-12 science education. National Science Teachers Association.

BACKPACK SCIENCES

DeBoer, G. E. (2014). A history of ideas in science education: Implications for practice. Routledge.

Hohmann, M., & Weikart, D. (1995). Educating young children: Active learning practices for preschool and child care programs. High/Scope Press.

Klopfer, E. D., Osterweil, S., Groff, J., & Haas, J. (2009). Using the technology of today, in the classroom today: The instructional power of digital games, social networking, simulations and how teachers can leverage them. The Education Arcade.

Lee, O., Deaktor, R., Enders, C., Lambert, J., & Robinson, D. (2007). Science education and student diversity: Synthesis and research agenda. Cambridge University Press.
National Science Teaching Association. (2019). Resources for teaching science through real-world issues.

Lillard, A. S., & Else-Quest, N. (2006). Evaluating Montessori education. Science, 313(5795), 1893-1894.

National Research Council. (2012). A framework for K-12 science education: Practices, crosscutting concepts, and core ideas. National Academies Press.

Osborne, J., Simon, S., & Collins, S. (2003). Attitudes towards science: A review of the literature and its implications. International journal of science education, 25(9), 1049-1079.

Science Education Resource Center at Carleton College. (n.d.). Using Issues to Teach Science.

Settlage, J., & Southerland, S. A. (Eds.). (2008). Picture-perfect science lessons: Using children's books to guide inquiry, K-4. NSTA Press.

Smetana, L. K., Bell, R. L., & Jansen, J. (2017). Preservice elementary teachers' connections between children's literature and science. Journal of Elementary Science Education, 29(3), 16-26.

TeachEngineering. (n.d.). What Is an Engineer?

Chapter 07

"Montessori, Technology, and the Future of Education" by Angeline Lillard (2012) is a scholarly article that explores the potential benefits and drawbacks of technology in Montessori education.

"The Montessori Method and the Montessori App: A Comparative Content Analysis of Science Education Materials" by Jane Y. Lee and Annahita Ball (2016) is another scholarly article that compares traditional Montessori science materials to a Montessori-based science app.

The Montessori Foundation website (montessori.org) offers a range of resources on Montessori education, including articles, videos, and research studies.

The American Montessori Society website (amshq.org) also provides resources on Montessori education, including articles, books, and research studies.

Chapter 08

Association Montessori Internationale (AMI). (2018). The Montessori approach to education. Retrieved from https://montessori-ami.org/about-montessori/montessori-approach-education.

Chawla, L. (1999). Life paths into effective environmental action. The Journal of Environmental Education, 31(1), 15-26.

Lillard, A. S., & Else-Quest, N. (2006). Evaluating Montessori education. Science, 313(5795), 1893-1894. doi: 10.1126/science.1132362

Lillard, A. S. (2012). Preschool children's development in classic Montessori, supplemented Montessori, and conventional programs. Journal of School Psychology, 50(3), 379-401.

Lillard, A. S. (2013). Playful learning and Montessori education. American Journal of Play, 6(1), 80-98.

Lillard, A. S. (2017). Montessori: The science behind the genius. Oxford University Press.

Montessori, M. (1912). The Montessori Method: Scientific Pedagogy as Applied to Child Education in "The Children's Houses". Frederick A. Stokes Company.

Montessori, M. (1917). Spontaneous Activity in Education. The Montessori Series.

"Observation: The Key to Understanding Your Child" by Maria Montessori (https://amshq.org/Family-Resources/Montessori-Articles/Observation-The-Key-to-Understanding-Your-Child)

"Observation in the Montessori Environment" by The Montessori Notebook (https://www.themontessorinotebook.com/observation-in-the-montessori-environment/)

"Observation in the Montessori Classroom" by Montitute (https://montitute.com/observation-in-the-montessori-classroom/)

"Observation: The Heart of the Montessori Method" by Montessori Training (https://montessoritraining.blogspot.com/2010/01/observation-heart-of-montessori-method.html)

"Observation in Montessori Education" by Baan Dek (https://baandek.org/posts/observation-in-montessori-education/)

Rathunde, K. (2016). A review of research on Montessori education. Montessori Life, 28(4), 36-45.

Sander, S., & Sanders, E. (2018). The impact of Montessori education on primary school children's mathematics achievement: A longitudinal study. Journal of Research in Childhood Education, 32(2), 251-263. doi: 10.1080/02568543.2018.1447332

Schmidt, M., Harms, U., & Dueber, W. (2017). Impact of Montessori education on children's academic achievement, motivation and self-concept. International Journal of STEM Education, 4(1), 16. doi: 10.1186/s40594-017-0073-3

Standing, E. M. (1957). Maria Montessori: Her life and work. New American Library.

BACKPACK SCIENCES

"The Montessori Observer" by Montessori Education Institute of the Pacific Northwest (https://www.montessori-nw.org/sites/default/files/montessori_observer_2017.pdf)

"The Power of Observation in the Montessori Classroom" by Age of Montessori (https://ageofmontessori.org/the-power-of-observation-in-the-montessori-classroom/)

"Using Observation to Enhance Montessori Education" by Montessori Education (https://montessori-ami.org/resource-library/using-observation-enhance-montessori-education)

Weil, M. J., & Lillard, A. S. (2013). Preschool children's attention to ecological stimuli: Little evidence for the prepared environment. Frontiers in Psychology, 4, 925. doi: 10.3389/fpsyg.2013.00925

Chapter 09

Bradshaw, C. P., Waasdorp, T. E., & Leaf, P. J. (2012). Effects of school-wide positive behavioral interventions and supports on child behavior problems. Pediatrics, 130(5), e1136-e1145.

Carol Dweck's Mindset: The New Psychology of Success (2006)
The Importance of Celebrating Student Achievements by TeachHub (https://www.teachhub.com/importance-celebrating-student-achievements)

Citizen Science: Science for Everyone by Caren Cooper (https://www.citizenscience.org/)

Citizen Science: How Ordinary People are Changing the Face of Discovery by Darlene Cavalier and Eric B. Kennedy (https://www.amazon.com/Citizen-Science-Ordinary-Changing-Discovery/dp/1937994224)

Cohen, R. M. (2019). Culturally Responsive Teaching in the Montessori Classroom. In T. Ryan (Ed.), Montessori Education: A Review of the Evidence Base (pp. 169-187). Springer. https://doi.org/10.1007/978-3-030-14165-2_9

Experimental Design in Science: Definition & Method by K. Fogle (https://study.com/academy/lesson/experimental-design-in-science-definition-method.html)

Growth Mindset: What You Need to Know by Mindset Works (https://www.mindsetworks.com/science/)

Incorporating Multimedia in Classroom Presentations by Center for Teaching and Learning, University of North Carolina at Chapel Hill (https://ctl.unc.edu/resources/publications/Incorporating-Multimedia-in-Classroom-Presentations/)

Lillard, A. S., & Else-Quest, N. (2006). Evaluating Montessori education. Science, 313(5795), 1893–1894.

Lillard, A. S., Heise, M. J., Richey, E. M., Tong, X., Hart, A. J., & Bray, P. M. (2017). Montessori preschool elevates and equalizes child outcomes: A longitudinal study. Frontiers in psychology, 8, 1783. (https://doi.org/10.3389/fpsyg.2017.01783)

National Science Teachers Association. (2014). The Importance of Building Classroom Community in Science Class. Retrieved from (https://www.nsta.org/blog/importance-building-classroom-community-science-class)

National Science Teachers Association (NSTA) (https://www.nsta.org/)

Rathunde, K., & Csikszentmihalyi, M. (2005). Middle school students' motivation and quality of experience: A comparison of Montessori and traditional school environments. American Journal of Education, 111(3), 341–371.

Science Olympiad official website (https://www.soinc.org/)

Service Learning in Science: A Natural Connection by Sarah E. Anderson and Joseph R. Boyle (https://www.nsta.org/publications/news/story.aspx?id=53184)

ABOUT THE AUTHOR

Before finishing her undergraduate studies at the University of Colorado at Boulder, Jackie Grundberg, had already dipped her toes in studying animal behavior, worked at the City of Boulder Open Space in Colorado, and spent summers studying abroad in a wildlife management program in Kenya and completing an internship at Mystic Marinelife Aquarium in Connecticut.

After university, she worked full-time with the City of Boulder Open Space. She then worked for the Nature Conservancy in Long Island, New York. Eventually, she took a research position with the University of San Francisco, where she worked in a remote camp in Cameroon studying black-capped hornbills.

When Jackie moved to Japan to live on a U.S. military base, she utilized her biology skills and pursued her secondary teaching credentials to teach high school advanced placement biology. When she returned to the U.S., she discovered the Montessori philosophy. She earned her middle school and elementary state teaching credentials, a Master's in Educational Technology, and American Montessori Society (AMS) 6- to 12-year-old elementary credentials. She taught in the Montessori elementary classroom for 12 years.

In 2017, she began working at the Center for Guided Montessori Studies (CGMS), helping new educators earn their Montessori teaching credentials. She now works as an instructional guide, practicum advisor, and field consultant.

In 2019, she started her business, Backpack Sciences, where she helps educators overcome the pressure of teaching hands-on science by providing ready-to-go science lessons and activities. At the publishing date, she has helped over 30,000 teachers.

For more information about Backpack Sciences membership: www.backpacksciences.com

Jackie lives in Houston, Texas, and loves to travel around the world with her husband, Brent, and two children. Both children were in Montessori schools from the toddler program until 8th grade. The house is energized by Winston, their labradoodle, and the newly adopted stray cat, Cat/Oliver/Spike. (We haven't figured out a name yet.)

You can follow Jackie's adventures and tips for teaching science on social media (@backpacksciences), and even better, take one of her workshops or become a member.

ACKNOWLEDGMENTS

Reflecting on my journey, I'm humbled by the unexpected turns life has taken me on. Despite the twists and challenges, I'm profoundly grateful for the encouragement of those who have stood by me through it all. Their belief in me has been the driving force behind every triumph and transformation.

This book is dedicated to my husband, Brent, whose unwavering support, practical wisdom, and belief in me have been my rock throughout this journey. Your guidance, patience, and encouragement have been a constant source of strength, shaping not only my path but encouraging me to pursue my dreams. I am endlessly grateful for your presence by my side.

And to my dear children, Jonathan and Katherine, your curiosity and love infuse these pages with meaning and purpose. Your love for science, laughter, and enthusiasm remind me every day of what truly matters, inspiring me to continue to support others to create a brighter future. Thank you for being the heart of my inspiration.

To my beloved parents, though you are no longer with me on this earthly journey, your unwavering passion, sacrifice, hard work, and determination continue to inspire me every day. I owe everything I am to the values and lessons you instilled in me. As immigrants to the United States with only a few hundred dollars in your pockets, you showed us the true meaning of resilience and success. While I wish you could have held a copy of this book in your hands, your spirit and legacy live on in every word written.

To my childhood friend Melanie Lau, whose persistent encouragement and belief in me planted the seed of writing a book years ago. Despite my initial reluctance and laughter at the idea, your unwavering belief in me and repeated mentions of writing a book gradually began to resonate. With your own experience as an author, you inspired and guided me, showing me the ropes and demonstrating that writing a book was not only feasible but also immensely rewarding. Your unwavering persistence, invaluable assistance with the logistics of publishing, and constant words of encouragement have been instrumental in bringing this book to fruition. Your own success in writing books has been a profound inspiration to me, and I am deeply grateful for your friendship and support throughout this journey.

I thank all of the incredible Montessori educators and instructors I've worked with, collaborated with, or given me the chance to teach or speak at a conference. Each of you has helped me develop into a better Montessorian.

Seemi Abdullah, Laura Alexander, Aimee Allen, Rose Armand, Anya Bartlett, Michele Berrigan, Geoffrey Bishop, Kitty Bravo, Kristen Richter Brown, Diana Butler, Sonya Cary, Mayeen Clayton, Mary Clemer, Jane Collins, Bharathi Dhanakoti, Ewelina Donatti, Reinaldo Donatti, Michael Dorer, Sharon Dunn, Jo Ebisujima, Spramani Elaun, Yuliya Fruman, Christine Garcharna, Mary Glenn, Frieda Hammett, Heidi Harbaugh, Jana Morgan Herman, Sherry Herron, Shane Hickson, Teresa Hoggatt, Liz Hoyer, Katherine Kabral, Roshi Karunatileka, Jennifer Kilgore, Rachel Kincaid, Gabriela Knaudt, Claudia Krikorian, Andrew Kutt, Sharon Ledesma, Claudia Mann, Lorna McGrath, Stefanie Melo, Tammy Oesting, Sunita Pailoor, Stephanie Pullman, Pam Purdue, Angela Rambukwella, Letty Rising, Tanya Ryskind, Mary Schneider, Tim Seldin, Joanne Shango, Kay Shields, Munir Shivji, Priscilla Spears, Kristan Taylor, Amy Taylor-Concepcion, Kathleen Wallace, Heather White, Ann Winkler, Jonathan Wolff, and Carol Zielke.

In my journey's chapters, a heartfelt thank you to Stu McClaren, my incredible business mentor, whose wisdom and insights sparked the creation of my business, and show me how to share my talent and foster success with my community. His community is now my supportive community and has led me to transformative experiences like 29029, reshaping my path with a newfound purpose.

To Arvin Trinidad, my exceptional assistant and layout designer, whose talent, dedication, and meticulous attention to detail have played a pivotal role in bringing this book to life. Beyond his invaluable contributions to numerous Backpack Sciences projects, Arvin's expertise has been instrumental in crafting the visual aesthetic that makes this book both beautiful and seamlessly flowing. I am deeply grateful for his professionalism, creativity, and unwavering commitment to excellence throughout this collaborative journey.

Special thanks to the Tribe/Connect/Mastermind community led by Stu McClaren, whose unwavering encouragement and shared journey toward writing a book provided the final push I needed to embark on this endeavor. Your collective dedication to growth and support has been a constant source of inspiration. Additionally, heartfelt appreciation goes out to my dear friends, whose presence and unwavering support have kept me grounded and happy throughout life's ups and downs. Your friendship means the world to me, and I am deeply grateful for each of you.

SPECIAL REQUEST

WOULD YOU LIKE TO HELP ANOTHER EDUCATOR?

I hope that you use *"Teaching Montessori Science"* as a guide to increase engagement with children and to gain practical tools and strategies to teach hands-on science confidently.

If you have found this book valuable, please take a brief moment and leave an honest review.

https://www.amazon.com/review/create-review/?ie=UT-F8&channel=awUDPv3&asin=B0CY7B2Q1S

Thank you!

Seeking extra support?

Download the
Free Skeletal System Lesson Plan

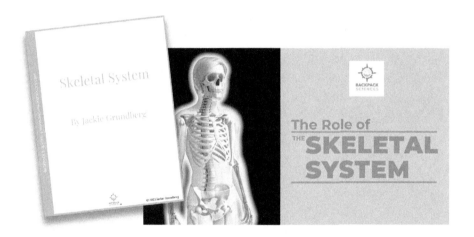

https://www.backpacksciences.com/skeletal

What's included:

- A full written lesson plan aligned with
  NGSS and Common Core standards
- A video of Jackie teaching the lesson presentation cards
- Montessori 5 part cards
- Hands-on, multidisciplinary follow-up activities

www.backpacksciences.com

Printed in Great Britain
by Amazon

43956431R00129